Leckie×Leckie
Scotland's leading educational publishers

N5 BIOLOGY STUDENT BOOK

National 5
BIOLOGY
STUDENT BOOK

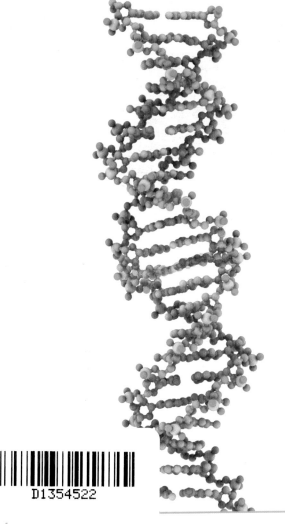

D1354522

Claire Bocian • Dianne Forrest
Bryony Smith

© 2018 Leckie & Leckie Ltd

001/03102018

10 9 8 7 6 5 4 3 2 1

ISBN 9780008282073

Published by
Leckie & Leckie Ltd
An imprint of HarperCollins Publishers
Westerhill Road, Bishopbriggs, Glasgow, G64 2QT
T: 0844 576 8126 F: 0844 576 8131
leckieandleckie@harpercollins.co.uk   www.leckieandleckie.co.uk

Special thanks to
Jouve (layout); Jess White (proofread); Janice McNeillie (project management);
Graham Moffat (peer review)

A CIP Catalogue record for this book is available from the British Library.

Acknowledgements
Leckie & Leckie would like to thank the following copyright holders for permission to reproduce their material:

Cover image © Lculig

Assessment questions Unit 1 B8, B9; Unit 2 A4, A9, A10, B5, B7; Unit 3 B3 adapted or taken from past examination papers © Scottish Qualifications Authority. The solutions do not emanate from the SQA.

Unit 1 chapter break: Dimarion; 1.1: Jupiterimages; 1.2: Duncan Smith; 1.12: THOMAS DEERINCK, NCMIR/SCIENCE PHOTO LIBRARY; 3.4: A. BARRINGTON BROWN/SCIENCE PHOTO LIBRARY; 4.9: MARTYN F. CHILLMAID/SCIENCE PHOTO LIBRARY; Unit 2 chapter break: Jiri Vaclavek; 7.7: Roz Woodward; 7.11: SCIENCE PHOTO LIBRARY; 7.12: J.C. REVY, ISM/SCIENCE PHOTO LIBRARY; 8.1: DR COLIN CHUMBLEY/SCIENCE PHOTO LIBRARY; 8.5: MEHAU KULYK/SCIENCE PHOTO LIBRARY; 8.10: ALAIN POL, ISM/SCIENCE PHOTO LIBRARY; 8.13: Jupiterimages; 8.18: CNRI/SCIENCE PHOTO LIBRARY; 8.26: Jeffrey Hamilton; 9.1: George Doyle & Ciaran Griffin; 9.12: POWER AND SYRED/SCIENCE PHOTO LIBRARY; 9.18: DR JEREMY BURGESS/SCIENCE PHOTO LIBRARY; 10.3: Comstock Images; 10.22: SCIENCE PHOTO LIBRARY; 11.4: POWER AND SYRED/SCIENCE PHOTO LIBRARY; 11.14: DR KEITH WHEELER/SCIENCE PHOTO LIBRARY; 11.19: DR KEITH WHEELER/SCIENCE PHOTO LIBRARY; 12.23: SCIENCE PHOTO LIBRARY; 12.34: EYE OF SCIENCE/SCIENCE PHOTO LIBRARY; Unit 3 chapter break: SiberianLena; 15.8: SHEILA TERRY/SCIENCE PHOTO LIBRARY; 15.9: DR. STANLEY FLEGLER, VISUALS UNLIMITED/SCIENCE PHOTO LIBRARY; 16.3: CORDELIA MOLLOY/SCIENCE PHOTO LIBRARY; 16.9: SCIENCE PHOTO LIBRARY; 16.12: POWER AND SYRED/SCIENCE PHOTO LIBRARY; 18.5: NIGEL CATTLIN/SCIENCE PHOTO LIBRARY; 19.1: MICHAEL W. TWEEDIE/SCIENCE PHOTO LIBRARY; 19.3: DR. STANLEY FLEGLER, VISUALS UNLIMITED/SCIENCE PHOTO LIBRARY; 19.9: Mr Green

Whilst every effort has been made to trace the copyright holders, in cases where this has been unsuccessful, or if any have inadvertently been overlooked, the Publishers would gladly receive any information enabling them to rectify any error or omission at the first opportunity.

Printed and bound by Grafica Veneta S.p.a, Italy

CONTENTS

Area of study 2 – MULTICELLULAR ORGANISMS

CONTENTS

Activities answers and review question answers at
https://collins.co.uk/pages/scottish-curriculum-free-resources

Introduction

Welcome to the National 5 Biology Student Book!

This book covers all of the skills, knowledge and understanding included in the National 5 Course, following the published National 5 Course Specification.

It has been designed to help you pass National 5 Biology and is suitable for students who are currently following a National 5 course. Entry to the Course is at the discretion of a centre. However, we have assumed some prior skills and knowledge that would be gained from achievement of the fourth curriculum level or the National 4 Biology Course. Students may also progress from relevant chemistry, environmental science, physics or science courses.

The book has three main areas of study which are subdivided into chapters. Each chapter covers the mandatory knowledge with integrated skills, to cover all of the key areas within the course specification. The book has a number of features to assist your learning.

Features

YOU SHOULD ALREADY KNOW:

Each chapter starts with a summary of the assumed prior knowledge. This should help to give you a background to the ideas that will be explored within the chapter.

You should already know:

- Animal and plant species depend on each other, and the removal of a species will impact upon other species in the ecosystem.

LEARNING INTENTIONS

Each chapter opens with a list of topics covered in the chapter, and gives you a good idea as to what you should be able to do when you have worked your way through the chapter.

Learning intentions

- Use the correct terminology to describe an ecosystem.

HINTS

Where appropriate, Hints are given to help give extra support and assist your learning.

> 🔍 **Hint**
>
> The term 'nucleus' describes a single nucleus. If there is more than one nucleus, we say 'nuclei'.
>
> The term 'mitochondrion' describes a single mitochondrion. If there is more than one mitochondrion, we say 'mitochondria'.

MAKE THE LINK

There are two types of Make the link featured in the book. Make the link – Biology shows how one topic links to another area in the National 5 course. Other Make the links highlight links between subjects and help to show how the knowledge and skills gained from studying biology can be transferred to other areas.

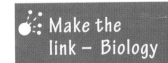

Make the link – Biology

Within Area of study 1 you will learn more about what happens in each of the cell structures. The role of ribosomes in protein synthesis (Chapter 3) and the role of mitochondria in aerobic respiration (Chapter 6) will be examined in greater detail.

BIOLOGY IN CONTEXT

Throughout the book, examples are given that show the real-life applications of the topics in the course.

Biology in context

After fertilisation the zygote divides many times. When it has formed a ball of 64 cells it is known as an embryo. The embryo implants into the wall of the uterus and begins to develop. The embryo receives nutrients from its mother through the uterus lining. After 3 months it has developed into a foetus and receives nutrients from the placenta

ACTIVITIES

These are sets of questions covering knowledge and skills and active learning tasks to work on individually, with a partner or in a group.

GO! Activities

Activity 1.1.1 Working individually

1. Make drawings of the ultrastructure of each of the four different cell types. Add arrows to identify each of the different structures in each of the cells. Using a separate piece of paper, cut out a set of labels and practise identifying the different structures in each of your drawings.

THINKING ABOUT EXPERIMENTS

Throughout the textbook you will come across the 'Thinking about experiments' star. These highlight the seven techniques that you should be familiar with. You can practise your understanding of each technique by following the page references to the specific examples given in the 'Thinking about experiments' chapter at the end of the book.

Design features of the experiment:

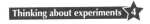
Thinking about experiments ⭐ 4

1. **Give** a reason why the enzyme is immobilised on the jelly bead.
2. **Predict** the time taken for sugar to be present if the experiment had been carried out at 1°C. **Explain** your prediction.
3. **State** the control for this experiment.
4. **Give** two variables, not already mentioned, that would have to be kept the same to make sure that the experiment was valid.
5. **Explain** how the milk could be kept at the temperature required for each experiment.
6. **Explain** how the results could be made more reliable.
7. **Predict** the time taken for sugar to be present if this experiment had been repeated at 60°C.

SUCCESS CRITERIA

At the end of specific sections within each chapter there is a summary of learning statements showing what you should be able to do when you complete the chapter. You should use them to help identify where you are with your learning and the next steps needed to make any improvements.

I can:

- Name and identify the following structures found in the ultrastructure of an animal cell: nucleus, cell membrane, cytoplasm, mitochondria and ribosomes.

ASSESSMENT

Review questions are provided at the end of each area of study.

These take a similar format to the question papers you will sit in the National 5 exam.

ANSWERS

Answers to knowledge and understanding activities and the review questions are provided online at https://collins.co.uk/pages/scottish-curriculum-free-resources.

Course award

The grade awarded is based on the total marks achieved across all course assessment components.

Course assessment

The course assessment has two components: the question paper and the assignment.

Course assessment structure

COMPONENT 1 – THE QUESTION PAPER

The question paper has two sections.

Section 1, contains multiple-choice questions and has 25 marks.

Section 2, contains structured and extended response questions and has 75 marks.

Total of 100 marks (80% of the total marks for external assessment).

Marks are distributed proportionally across the course content.

The majority of marks are awarded for demonstrating and applying knowledge and understanding. The other marks are awarded for applying scientific inquiry, scientific analytical thinking, problem-solving skills and the impact of applications of biology on society and the environment.

The question paper is 2 hours and 30 minutes in duration. It is set and marked by SQA, and conducted in centres under conditions specified for external examinations by SQA.

Specimen question papers for National 5 courses are published on the SQA website.

In addition to the knowledge and skills in each of the key areas of the Course, you should have knowledge of the following pieces of apparatus and techniques.

Apparatus: beaker, balance, measuring cylinder, dropper/pipette, test tube/boiling tube, thermometer, funnel, syringe, timer/stopwatch, microscope, Petri dish, quadrat, pitfall trap, light/moisture meter, water bath.

Experimental techniques: measuring enzyme activity, using a respirometer, measuring transpiration using a potometer, measuring abiotic factors, measuring the distribution of a species, using a transect line, measuring the rate of photosynthesis.

The apparatus and experimental techniques identified here are referenced in the text of the book and exemplified in the final chapter, 'Thinking about experiments'. Knowledge of both can be assessed in the question paper.

COMPONENT 2 – THE ASSIGNMENT

The purpose of the assignment is to assess the application of skills of scientific inquiry and related biology knowledge and understanding.

Candidates are required to:

- choose, with support, a relevant topic in biology
- devise an appropriate aim
- give an account of biology relevant to the aim
- plan and carry out experimental work/fieldwork to generate data relevant to the aim
- process and present the experimental/fieldwork data
- compare data/information from internet/literature research with the experimental/ fieldwork data
- draw a conclusion
- evaluate the experimental/fieldwork procedure
- communicate the findings in a report

The assignment has two stages: research (conducted under some supervision and control) and a report (conducted under a high degree of supervision and control). Part of the research requires you to plan and carry out an experiment and collect results. You could think about using one of the techniques outlined in the 'Thinking about experiments' chapter as your method for this. It is not a requirement to use one of these but it is certainly worth considering.

It is recommended that no more than 8 hours is spent on the whole assignment. A maximum of 1 hour and 30 minutes is allowed for the report stage. There is no word count.

The assignment is worth 20 marks, which will be scaled to 25 to calculate the overall grade.

Sections of Assignment	Marks
Aim	1
Underlying Biology	3
Data collection and handling	6
Graphical Presentation	4
Analysis	1
Conclusion	1
Evaluation	2
Structure	2
Total Marks	**20**

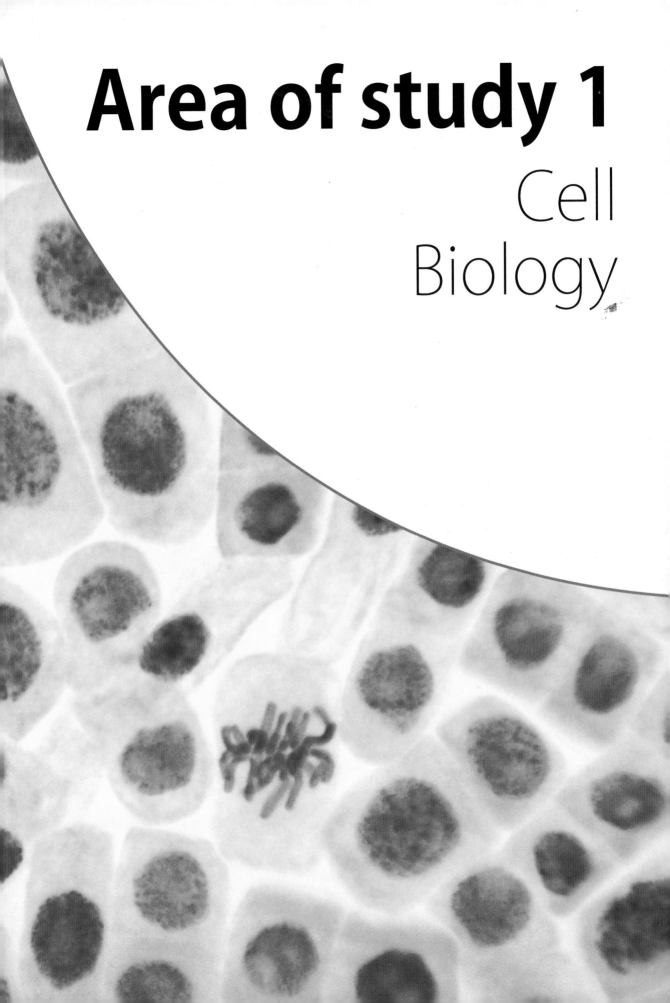

Area of study 1
Cell
Biology

1 Cell structure

You should already know:

- Cells are the basic units of all living things.
- Animal and plant cells have a nucleus, cell membrane and cytoplasm.
- Plant cells also have a cell wall, sap vacuole and chloroplasts.
- The cell membrane controls what enters and leaves the cell.
- The nucleus contains genetic information and controls all cell activities.
- The cytoplasm is the site of chemical reactions.
- The cell wall is freely permeable and is involved in support of the plant cell.
- The sap vacuole helps to keep the shape of the cell.
- The chloroplasts are the site of photosynthesis.

Learning intentions

- Name the structures found in the ultrastructure of an animal cell.
- Identify in a diagram the structures found in the ultrastructure of an animal cell.
- State the functions of the structures found in the ultrastructure of an animal cell.
- Name the structures found in the ultrastructure of a plant cell.
- Identify in a diagram the structures found in the ultrastructure of a plant cell.
- State the functions of the structures found in the ultrastructure of a plant cell.
- Name the structural material of which a plant cell wall is composed.
- Name the structures found in the ultrastructure of a fungal cell.
- Identify in a diagram the structures found in the ultrastructure of a fungal cell.
- State the functions of the structures found in the ultrastructure of a fungal cell.
- Describe the structural similarities and differences between different cell types.
- Compare the cell wall structure in plant, fungal and bacterial cells.
- Name the structures found in the ultrastructure of a bacterial cell.
- Identify in a diagram the structures found in the ultrastructure of a bacterial cell.
- State the functions of the structures found in the ultrastructure of a bacterial cell.
- Calculate cell sizes based on their magnification.

Basic cell structure

The basic structures of animal and plant cells can be viewed using a light microscope (fig 1.1). Simple biological line drawings can be made of these cells.

Cheek epithelial cells are a good example of basic animal cells. A swab can be taken from the inside of the cheek and examined under the microscope. Biological stains are used on cells to allow the different cell structures to be clearly viewed. Using a light microscope allows some of the organelles that are present in animal cells to be seen.

Fig 1.1 *Using a light microscope*

Figure 1.2 shows the basic structures visible in an animal cell using a light microscope, e.g. cheek epithelial cells.

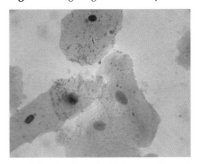

The names of each structure found in a basic animal cell are shown in the simple illustration below (fig 1.3).

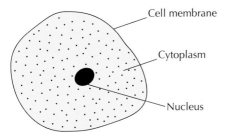

Fig 1.3 *Basic animal cell*

Fig 1.2 *Cheek epithelial cells viewed under a light microscope*

Cells from onion, rhubarb and *Elodea* (Canadian pondweed) are just some examples of plant cells that can be viewed under the light microscope.

Onion epidermal cells can be stained with iodine and viewed using a light microscope (fig 1.4).

Fig 1.4 *Plant cells viewed under a light microscope*

Like an animal cell, a plant cell has a nucleus, cytoplasm and cell membrane. Also visible in figure 1.4 is the cell wall. Chloroplasts are only visible when green parts of a plant are viewed under the microscope.

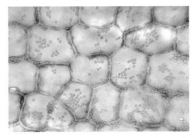

In figure 1.5, the chloroplasts are clearly visible when *Elodea* is viewed under the microscope.

Fig 1.5 *Elodea (Canadian pondweed) cells under a light microscope*

The line drawing (fig 1.6) shows the names of each structure found in a plant cell.

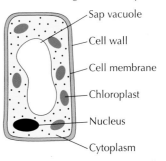

Fig 1.6 *The parts of a plant cell*

Fig 1.7 *Light microscope*

Fig 1.8 *Electron microscope*

As a light microscope (fig 1.7) has limited magnification, an electron microscope (fig 1.8) is used to view cells in greater detail. With this microscope other organelles present in the cytoplasm become visible. What can be seen within the cell is described as the **cell ultrastructure**.

Ultrastructure of an animal cell

In addition to the **nucleus**, **cell membrane** and **cytoplasm**, other structures can be seen using the electron microscope.

The illustration in figure 1.9 shows the ultrastructure of an animal cell as revealed by an electron microscope. In addition to the nucleus, cytoplasm and cell membrane, there are two other structures. The **mitochondrion** is the site of aerobic respiration, and the **ribosomes** are where proteins are synthesised.

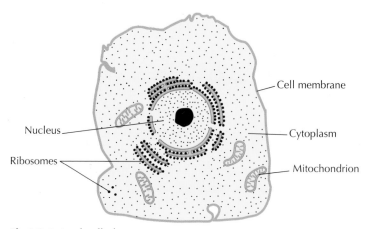

Fig 1.9 *Animal cell ultrastructure*

> ### 🔎 Hint
>
> The term 'nucleus' describes a single nucleus. If there is more than one nucleus, we say 'nuclei'.
>
> The term 'mitochondrion' describes a single mitochondrion. If there is more than one mitochondrion, we say 'mitochondria'.

Cells can be thought of as mini factories with jobs being carried out in specialised areas. Each structure is designed in a way that suits its function. Specific organelles carry out their function to ensure the survival of the cell.

Ultrastructure of a plant cell

The structures normally visible in a plant cell under the light microscope are: the nucleus, cell membrane, cytoplasm, **cell wall**, **sap vacuole** and **chloroplasts**. Using an electron microscope allows you to see the chloroplast in more detail. Ribosomes and mitochondria are also visible and have the same function as in animal cells. The cell wall in a plant cell is made of a structural carbohydrate called **cellulose**.

The illustration below shows the organelles present in a plant cell (fig 1.10).

Ultrastructure of a fungal cell

Yeast is an example of a unicellular fungus. The structure of a yeast cell is very similar to an animal cell. It has a nucleus, cell membrane and cytoplasm (fig 1.11). The only difference between the animal and yeast cell is the presence of a vacuole and cell wall in the yeast cell. All structures found in animal, plant and fungal cells perform the same functions. The presence of a cell wall is common to both fungal and plant cells. However, the walls are structurally different and composed of different materials.

> ### ⚗ Make the link – Biology
>
> Within Area of study 1 you will learn more about what happens in each of the cell structures. The role of ribosomes in protein synthesis (Chapter 3) and the role of mitochondria in aerobic respiration (Chapter 6) will be examined in greater detail.

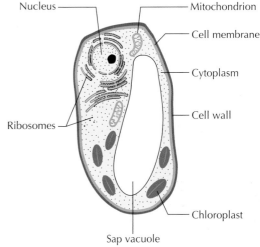

Fig 1.10 *The structure of a plant cell*

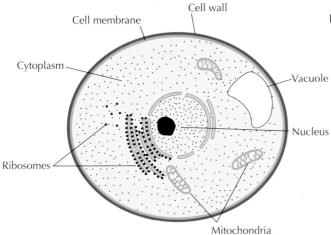

Fig 1.11 *The organelles present in a yeast cell*

> ### ⚗ Make the link – Biology
>
> Within Area of study 3 you will learn more about the process of photosynthesis that takes place inside chloroplasts (Chapter 16).

There are a variety of different types of yeast. Different strains of yeast are used in the baking and brewing industries.

The illustration in figure 1.12 shows the organelles present in a yeast cell.

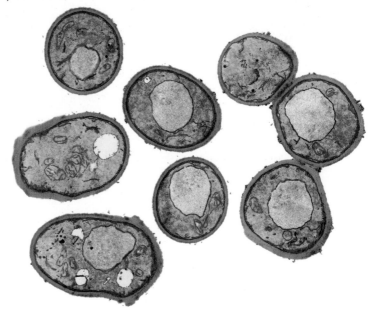

Fig 1.12 *Yeast cells in colour under a high-power microscope*

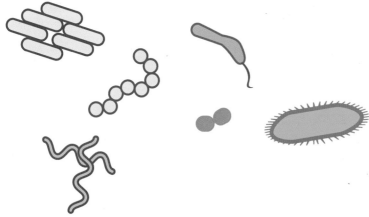

Fig 1.13 *The different forms bacterial cells can take*

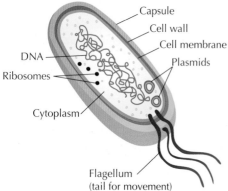

Capsule
Cell wall
Cell membrane
DNA
Plasmids
Ribosomes
Cytoplasm
Flagellum
(tail for movement)

Fig 1.14 *The structures present in a typical bacterial cell*

Ultrastructure of a bacterial cell

There are many different types of bacteria. Like the other cell types, the bacterial cell has a cell membrane, cell wall, cytoplasm and ribosomes.

The most obvious difference between a bacterial cell and the other cell types is the lack of a nucleus. Also absent from the bacterial cell are organelles such as mitochondria, a vacuole and chloroplasts. Instead of a nucleus the bacterial cell contains DNA as well as circular **plasmids** that also contain DNA. The cell wall structure of the bacterial cell is again different to those found in plant and yeast cells. Although all three cell types (plant, fungal and bacterial cells) have a cell wall, the chemical composition of all three is different. Only plant cell walls contain cellulose. Fungal and bacterial cell walls are made of different materials.

Figure 1.13 shows that bacterial cells come in many different shapes and sizes. Some bacteria are harmful, while others are useful.

The illustration in figure 1.14 shows the structures present in a typical bacterial cell.

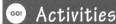 **Activities**

Activity 1.1.1 Working individually

1. Make drawings of the ultrastructure of each of the four different cell types. Add arrows to identify each of the different structures in each of the cells. Using a separate piece of paper, cut out a set of labels and practise identifying the different structures in each of your drawings.

2. Using a mini whiteboard, copy the table below. Add ticks to show which structures are found in each of the cell types.

Structures	Animal	Plant	Fungal	Bacterial
Cell membrane				
Cytoplasm				
Nucleus				
Cell wall				
Sap vacuole				
Chloroplast				
Mitochondria				
Ribosome				
Plasmid				

3. Answer the following questions:
   (a) Create the following lists:
      Name the cell structures that:
      (i) Animal, plant, fungal and bacterial cells all share.
      (ii) Animal and plant cells have in common.
      (iii) Animal and fungal cells have in common.
      (iv) Animal and bacterial cells have in common.
      (v) Plant and fungal cells have in common.
      (vi) Plant and bacterial cells have in common.
      (vii) Fungal and bacterial cells have in common.
   (b) Name the structural carbohydrate that plant cell walls are composed of.
   (c) Describe the differences between the structure of the cell walls in plant, fungal and bacterial cells.

Activity 1.1.2 Working in pairs

Using the drawings and labels from activity 1.1.1 play a cell structure lotto game. Each person picks two different cells types. Place all of the labels face down and take turns trying to find and correctly match up the label to its structure.

I can:

- Name and identify the following structures found in the ultrastructure of an animal cell: nucleus, cell membrane, cytoplasm, mitochondria and ribosomes.

- Name and identify the following structures found in the ultrastructure of a plant cell: nucleus, cell membrane, cytoplasm, cell wall, sap vacuole, chloroplast, mitochondria and ribosomes.

- Name cellulose as the structural material of which a plant cell wall is composed.

- Name and identify the following structures found in the ultrastructure of a fungal cell: nucleus, cell membrane, cytoplasm, cell wall, vacuole, mitochondria and ribosomes.

- State that a fungal cell has a cell wall, but it is different in structure to that of a plant cell wall.

- Name and identify the following structures found in the ultrastructure of a bacterial cell: cell membrane, cytoplasm, cell wall, plasmids and ribosomes.

- State that a bacterial cell has a cell wall, but it is different in structure to that of a plant cell wall or fungal cell wall.

- Describe fungal cell structure in terms of similarity to plant and animal cells but with a different cell wall structure.

- Describe bacterial cell structure in terms of the absence of organelles and a different cell wall structure to plant and fungal cells.

Cell functions

The table summarises the functions of all structures found in animal, plant, fungal and bacterial cells.

Structure	Function
Nucleus	Contains genetic information and controls all cell activities
Cell membrane	Selectively permeable; controls what enters and leaves the cell
Cytoplasm	Site of chemical reactions
Mitochondria (-ion)	Site of aerobic respiration (energy release)
Ribosome	Site of protein synthesis
Cell wall	Freely permeable; involved in support of the cell
Vacuole	Helps keep the shape of the cell
Chloroplast	Site of photosynthesis
Plasmid	Circular piece of DNA found in bacterial cells

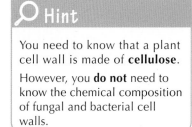

Hint

You need to know that a plant cell wall is made of **cellulose**.

However, you **do not** need to know the chemical composition of fungal and bacterial cell walls.

Activities

Activity 1.1.3 Working individually

On paper, draw a structure and function table. Complete the structures and functions using the table above. Repeat the process using a mini whiteboard. This time try to complete the table without using your notes.

Activity 1.1.4 Working in pairs

Each person should make nine small cards. One person writes down the names of the nine different cell structures. The other person writes down the functions of the nine structures. Work together to match up each structure with the correct function.

Activity 1.1.5 Working in groups

From memory, produce four different mini-posters to show the ultrastructure of each of the cell types – animal, plant, fungal and bacterial. Label all of the cell structures and describe their functions.

I can:

- State the functions of the structures (nucleus, cell membrane, cytoplasm, mitochondria and ribosomes) found in the ultrastructure of an animal cell.

- State the functions of the structures (nucleus, cell membrane, cytoplasm, cell wall, sap vacuole, chloroplast, mitochondria and ribosomes) found in the ultrastructure of a plant cell.

- State the functions of the structures (nucleus, cell membrane, cytoplasm, cell wall, vacuole, mitochondria and ribosomes) found in the ultrastructure of a fungal cell.

- State the functions of the structures (cell membrane, cytoplasm, cell wall, plasmids and ribosomes) found in the ultrastructure of a bacterial cell.

Cell sizes

Before the invention of the microscope, people had never seen cells. As technology advances, so do microscopes. This allows us to see the complexity of cells.

The size of an individual cell can be calculated based on the number of cells that are visible in the field of view. To calculate the size of an individual cell, we need to know the total magnification of the microscope. The total magnification is calculated by multiplying the eyepiece lens magnification by the objective lens magnification.

For example, if the eyepiece lens used was ×10, and the objective lens was ×4, the total magnification would be ×40.

As cells are microscopic they need to be measured in units smaller than millimetres, so they are measured in micrometres (also referred to as microns) and nanometres. 1 micrometre (μm) is equal to one thousandth of a millimetre. 1 nanometre is one thousandth of a micrometre. Using a light microscope we can see structures that can be measured in micrometres. With an electron microscope, structures measured in nanometres are visible. Figure 1.15 helps to put these sizes into context.

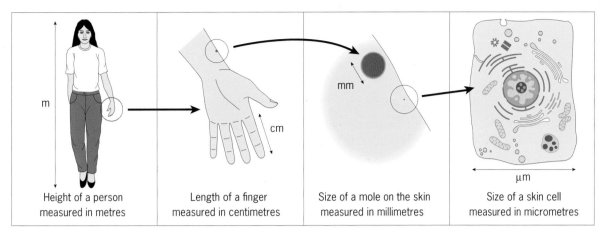

Fig 1.15 *The relationship between units of measurement*

Look at figure 1.16 as an example. The field of view is 2 mm in diameter.

To work out the length of an individual cell, the number of cells across the length of the field of view must be counted.

There are five cell lengths that cover the length of the field of view.

$$\frac{\text{Field of view}}{\text{Number of cells}} \times 1000 = \text{length of each cell.}$$

$$\frac{\text{Field of view}}{\text{Number of cells}} = \frac{2 \text{ mm}}{5 \text{ cells}} = 0.4 \text{ mm}$$

$0.4 \text{ mm} \times 1000 = 400 \text{ μm}$

Therefore the length of each cell is equal to 400 μm (fig 1.16).

The breadth of 10 cells covers the length of the field of view.

$$\frac{\text{Field of view}}{\text{Number of cells}} \times 1000 = \text{breadth of each cell.}$$

$$\frac{\text{Field of view}}{\text{Number of cells}} = \frac{2 \text{ mm}}{10 \text{ cells}} = 0.2 \text{ mm}$$

$0.2 \text{ mm} \times 1000 = 200 \text{ μm}$

Therefore the breadth of each cell is equal to 200 μm (fig 1.16).

Make the link – Biology

When you calculate areas of shapes in maths, you learn that the length is the longest side and the breadth, the shorter side. Cells have an irregular shape, but we still need to think of the longer part as the length and the shorter part the breadth.

Hint

Make sure that you can calculate the length and breadth of a single cell, taking into account the length/breadth of the field of view and the number of cells present.

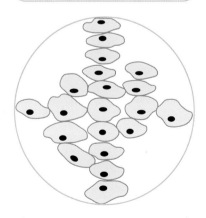

Field of view = 2 mm

Fig 1.16 *Measuring cell length using a microscope*

GO! Activities

Activity 1.1.6 Working in pairs

1. On a sheet of paper, each person should copy and complete the table below. Work together, using the formula to calculate the total magnification based on the same eyepiece being used each time.

Total magnification = eyepiece lens magnification × objective lens magnification

Eyepiece lens magnification	Objective lens magnification	Total magnification
× 10	× 4	
	× 10	
	× 40	

2. Now make your own diagrams of a 2 mm diameter field of view down a microscope. Insert between 2 and 20 cells (cells must touch each other and extend across the length of the field of view) then calculate the average length of a cell in microns. Show all of your calculations. Swap diagrams with a partner to check each other's calculations.

Activity 1.1.7 Working in groups

From memory, produce four different mini-posters to show the ultrastructure of each of the cell types – animal, plant, fungal and bacterial. Label all of the cell structures and describe their functions.

I can:

- Calculate total magnification using the formula:

  total magnification = eyepiece lens magnification × objective lens magnification

- Calculate the length and breadth of individual cells, shown in a drawing, when given the diameter of the field of view.

2 Transport across cell membranes

Learning intentions

- Describe the composition of the cell membrane.
- Describe how the structure of the membrane relates to its permeability.
- Define the term 'passive transport'.
- Define the term 'diffusion'.
- Explain how diffusion occurs across cell membranes.
- Define the term 'osmosis'.
- Explain how osmosis occurs across cell membranes.
- Describe the effects of osmosis on plant and animal cells.
- Define the term 'active transport'.
- Explain how active transport occurs across cell membranes.

The structure of the cell membrane

The cell membrane separates the cell contents from its external environment. It is made of **protein** and **phospholipid** molecules. The fluid mosaic model is commonly used to describe how proteins and phospholipids are arranged within the cell membrane.

The phospholipid molecules are found in a double or bi-layer that is in constant motion – this is the 'fluid' part. The flexibility of this double layer allows the cell to change in shape without causing damage to the cell.

The protein molecules are found in a patchy arrangement, spread throughout the phospholipid molecules – this gives the mosaic pattern. Some proteins pass through the width of the membrane and contain a channel, forming a pore. Other proteins are only partly embedded, and some lie on the surface (fig 2.1). The proteins have different functions.

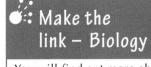

Make the link – Biology

You will find out more about the functions of proteins in Chapter 4 (Proteins).

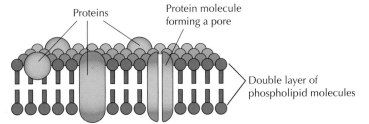

Proteins

Protein molecule forming a pore

Double layer of phospholipid molecules

Fig 2.1 *The structure of the cell membrane*

The channel-forming protein pores allow certain substances to enter and leave the cell. Owing to this property, cell membranes are described as being **selectively permeable**. Small soluble molecules can pass easily through the membrane. Large insoluble molecules must first be broken down before they can pass into or out of cells.

Hint

As cells are microscopic they are measured in micrometres (μm).
1 micrometre = 1 thousandth of a millimetre.

Activities

Activity 1.2.1 Working individually

Using a mini whiteboard, practise drawing and labelling the structure of the membrane. Make sure that you can correctly identify the proteins and phospholipids from memory.

Activity 1.2.2 Working in pairs

Described below is an experiment that is carried out to investigate the structure of the cell membrane. With your partner, read the information carefully and complete the task.

If the membrane is made of phospholipids or proteins, damage to either one of them should cause leakage of the cell contents. Red cabbage can be used to investigate the structure of the cell membrane. The vegetable is dark purple in colour, so any damage to the membrane results in the purple liquid leaking from the tissue. Proteins can be damaged at high temperatures, and phospholipids can be dissolved by alcohol.

The following experiment was carried out to investigate the composition of the membrane. 40 discs of red cabbage were cut and then rinsed in water until the water ran clear. They were then placed in test tubes under the following conditions:

(Continued)

Test tube 1: 10 cm³ of water and 10 discs of red cabbage left at room temperature for 15 minutes.

Test tube 2: 10 cm³ of water and 10 discs of red cabbage placed in a water bath at 90°C for 15 minutes.

Test tube 3: 10 cm³ of alcohol and 10 discs of red cabbage left at room temperature for 20 minutes.

Test tube 4: 10 cm³ of alcohol and 10 discs of red cabbage placed in a water bath at 90°C for 15 minutes.

Fig 2.2a *Discs of red cabbage at the start of the experiment*

Use figures 2.2a and b to make notes and draw conclusions on the structure of the cell membrane. Assess each test tube individually using the list below to help you to focus on each experiment in turn.

Examine – The before and after photographs for each test tube individually. Note that the cabbage has not disappeared; it has sunk to the bottom of each test tube. Examine the colour changes.

Observe – What do you notice about the colour of the water?

Discuss – Why has this change taken place? What condition (temperature/alcohol) caused the change? What part of the membrane (protein/phospholipid) has been affected?

Conclude – What effect did the condition have on the cell membrane? Why did this happen?

Fig 2.2b *Discs of red cabbage following various treatments*

I can:

- State that the cell membrane is made of phospholipids and proteins.
- Identify phospholipids and proteins on a diagram of the cell membrane.
- State that the cell membrane is selectively permeable.
- Explain that the membrane proteins have channels that allow substances to enter and leave the cell.

Movement of substances across the cell membrane

The movement of substances into and out of cells is essential for all living organisms. Movement of substances through the cell membrane occurs by three different transport methods:

(1) Diffusion

(2) Osmosis

(3) Active transport

Different concentrations of substances exist between cells and their environment. In order to understand the movement of substances into and out of cells, we need to understand the term 'concentration gradient'.

A **concentration gradient** can be described as a difference in concentration of a substance between two solutions or between two cells or cell/tissue and a solution. Within any liquid or gas, a concentration gradient can exist. This happens when there is an uneven distribution of molecules (fig 2.3).

In figure 2.3, the molecules move from where they are in a higher concentration to where they are in a lower concentration. This happens until the concentration of molecules is equal. The movement of these molecules is described in the process of **diffusion**.

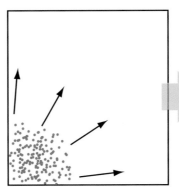
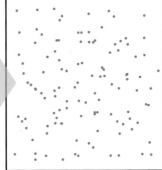

Fig 2.3 *Movement of molecules*

Passive transport

Passive transport is the movement of molecules of a substance down a concentration gradient and does not require additional energy. Small molecules can pass easily through a cell membrane by **passive transport**.

Diffusion

Diffusion is a process by which some molecules move into and out of cells. **Diffusion** is defined as **the movement of molecules down a concentration gradient from a higher to a lower concentration**. It is a passive process and does not require additional energy.

Make the link – Biology

Diffusion and osmosis are both examples of passive transport. They are examined in greater detail later in this chapter.

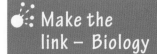

Make the link – Biology

In Chapter 6 of this Area of study you will learn about respiration in more detail. Diffusion is essential in respiration to allow the exchange of substances between cells and their environment.

In Area of study 2 (Chapters 12 and 13), you will learn about the exchange of materials in the bloodstream (page **190**) and gas exchange in the lungs (pages **194–195**). Diffusion is essential in both of these processes.

The importance of diffusion

Diffusion is essential to all living things. All cells need to take in food (e.g. glucose and amino acids) and oxygen. These can be used in respiration to provide energy needed by the cell to carry out all of its activities. Waste substances such as carbon dioxide can be removed from cells by diffusion.

Diffusion in action

Diffusion can only occur when there is a difference in concentration of a substance. The difference in concentration of a substance creates the concentration gradient. Molecules of a substance diffuse down a concentration gradient so diffusion requires no energy. Diffusion is an example of passive transport.

Diffusion can be easily demonstrated in a simple experiment using potassium permanganate crystals. A crystal of potassium permanganate can be placed in a beaker of still water. The crystal has a deep purple colour and will dissolve in the water over a period of time. Eventually the water will be the same colour throughout (fig 2.4).

At the start of the experiment, a concentration gradient exists between the crystal and the water. The molecules of the crystal are in a high concentration – they are clustered together. Over time they spread throughout the water, moving down the concentration gradient until the concentration is equal. By the end of the experiment, the water is the same colour throughout. The molecules have moved passively by diffusion.

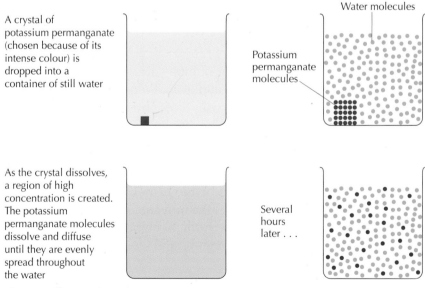

A crystal of potassium permanganate (chosen because of its intense colour) is dropped into a container of still water

As the crystal dissolves, a region of high concentration is created. The potassium permanganate molecules dissolve and diffuse until they are evenly spread throughout the water

Water molecules

Potassium permanganate molecules

Several hours later . . .

Fig 2.4 *Diffusion of potassium permanganate*

♈ Biology in context

Dissolving sugar in a cup of tea is an everyday example of diffusion. Even if you use a sugar cube and don't stir the sugar, it will still dissolve in the water, giving a sugary taste to every sip. This is diffusion in action. The sugar molecules move down the concentration gradient from where the sugar is in a higher concentration to where there is a lower concentration until they are evenly spread (fig 2.5).

 ● Water molecule

○ Sugar molecule

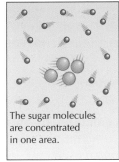

The sugar molecules are concentrated in one area.

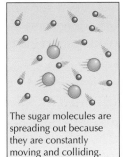

The sugar molecules are spreading out because they are constantly moving and colliding.

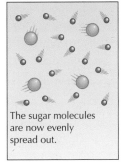

The sugar molecules are now evenly spread out.

Fig 2.5 *Diffusion of sugar molecules*

GO! Activities

Activity 1.2.3 Working individually

1. Write down full answers to the following questions:

    (a) Define the term 'passive transport'.

    (b) Define the term 'diffusion'.

    (c) Describe the relationship between diffusion and passive transport.

    (d) Using named substances, explain how the process of diffusion happens across cell membranes.

2. Make a simple drawing of a cup of tea. Repeat the same drawing two more times.

    In the first cup, add a sugar cube, as shown in figure 2.6, to show the particle arrangement of a solid.

    Using small circles to represent sugar particles, show what happens to the concentration of sugar in the cup over time. The third cup should show what happens once diffusion is complete.

Fig 2.6 *Particle arrangement of a sugar cube*

(Continued)

Activity 1.2.4 Working individually or in pairs

Make a simple diagram of a plant or animal cell. You should remember this from Chapter 1 (to practise your knowledge of cell ultrastructure make a more detailed drawing).

- Show the movement of the following substances: glucose, amino acids (both are examples of dissolved foods), oxygen and carbon dioxide.
- Create a key using different colours and/or shapes for each substance.
- Although all substances can move in and out of cells depending on the gradient, think carefully about which substances move into the cell as useful substances and which move out of the cell as waste substances and show this on the diagram.
- All of these substances move by the passive process of diffusion.
- Make sure that you show the substances moving from a higher concentration to a lower concentration. To show a higher concentration, make sure there are more molecules of the substance in comparison with the number of molecules of the same substance at a lower concentration.

I can:

- State that passive transport is the movement of molecules of a substance down a concentration gradient and does not require energy.
- State that different concentrations of substances exist between cells and their environment.
- Name diffusion and osmosis as examples of passive transport.
- State that diffusion is the movement of a substance from an area of high concentration to an area of low concentration until evenly spread.
- Describe diffusion in terms of concentration gradients.
- Name glucose, carbon dioxide, oxygen and amino acids as examples of substances that diffuse across cell membranes.
- Explain the importance of diffusion to organisms as being the means by which substances enter and leave cells by movement down the concentration gradient.

Osmosis

The diffusion of water through a membrane is called **osmosis**. This is defined as **the movement of water molecules from a higher water concentration to a lower water concentration through a selectively permeable membrane** (fig 2.7). Water molecules pass easily through a selectively permeable membrane, as they are very small.

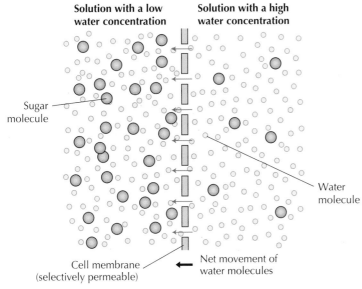

Fig 2.7 *Movement of water through a selectively permeable membrane*

The water molecules can pass freely through the membrane between the two solutions. Overall, there is a greater movement of water molecules from the solution with the higher water concentration to the solution with the lower water concentration. The water moves from where it is in a higher concentration to where it is in a lower concentration until its molecules are evenly spread.

The effects of osmosis on cells

Different concentrations of solutions affect cells in different ways. This can be demonstrated by some simple experiments using model cells.

Visking tubing is selectively permeable and can be used to represent the cell membrane.

Two Visking tubing bags are set up as model cells. Both contain a weak sugar solution. One is placed into a beaker containing a solution of pure water (beaker A) and the other into a beaker containing a strong sugar solution (beaker B) (fig 2.8).

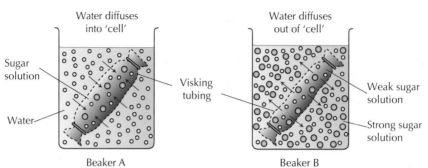

Fig 2.8 *Movement of water molecules in different solutions*

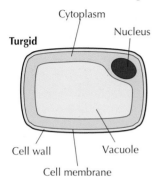

Fig 2.9 *A turgid plant cell*

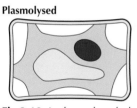

Fig 2.10 *A plasmolysed plant cell*

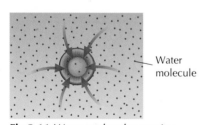

Fig 2.11 *Water molecules moving into a red blood cell by osmosis*

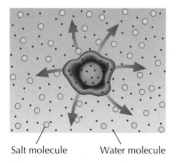

Fig 2.12 *Water molecules moving out of a red blood cell by osmosis*

From figure 2.8 we can see that in beaker A the water molecules move from the region of higher water concentration in the beaker of pure water into the Visking tubing bag where there is a lower water concentration. In beaker B the water molecules move out of the Visking tubing bag from a region of higher water concentration into the strong sugar solution where there is a lower water concentration.

Osmosis in plant cells

Plant cells contain a weak sugar solution inside their sap vacuole. When a plant cell is placed into pure water, it swells but does not burst. This is due to the strong cell wall. When water molecules enter the cell from the higher water concentration, the membrane is pushed up against the cell wall causing it to bulge. This can be seen using a microscope. Plant cells viewed under the microscope are described as being **turgid** (fig 2.9).

Plant cells placed into a concentrated sugar solution with a lower water concentration lose water to the solution. Water molecules move out from a region of higher water concentration inside the cell to a region of lower water concentration outside the cell. This causes the membrane to draw away from the cell wall. When viewed under the microscope, these cells are described as being **plasmolysed** (fig 2.10).

Osmosis in animal cells

Animal cells, like red blood cells, contain a weak salt solution. When a red blood cell is placed into a solution of pure water, a concentration gradient is created. Water molecules move from the higher water concentration outside the cell to the lower concentration inside the cell. As an animal cell has no cell wall to give it protection, the blood cell swells with water and eventually **bursts** (fig 2.11).

If a red blood cell is placed into a solution with a lower water concentration than the cell contents, water molecules will leave the cell and move into the solution. The water molecules move from a higher water concentration inside the cell to the lower water concentration outside the cell. When this happens the cell loses water, making it **shrink** (fig 2.12).

🔵GO! Activity

Activity 1.2.5 Working individually

1. Write down full answers to the following questions:

   (a) Define the term 'osmosis'.

   (b) Explain how the process of osmosis moves water molecules across cell membranes.

   (c) Describe the osmotic effects of the following:
   - transferring animal cells from a weak salt solution to a strong salt solution
   - transferring animal cells from a weak salt solution to a solution of pure water
   - transferring plant cells from a weak sugar solution to a strong sugar solution
   - transferring plant cells from a weak sugar solution to a solution of pure water

2. Read the information below then complete the task.

   The experiment described below was set up to investigate the effects of osmosis on plant tissue (fig 2.13).

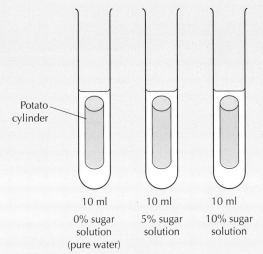

Potato cylinder

| 10 ml | 10 ml | 10 ml |
| 0% sugar solution (pure water) | 5% sugar solution | 10% sugar solution |

Fig 2.13 *Experiment to investigate the effects of osmosis on plant tissue*

   (a) Three boiling tubes were set up: one with 10 ml of pure water, one with 10 ml of 5% sugar solution and one with 10 ml of 10% sugar solution.

   (b) Three cylinders of potato were cut and blotted to remove excess water.

   (c) The mass of each cylinder was recorded before being placed into the different solutions.

   (d) The experiment was left set up for 2 hours.

   (e) The cylinders of potato were removed from the boiling tubes, blotted and re-weighed.

(Continued)

(f) The mass of the potato cylinders was recorded.

(g) The change in mass was calculated (difference between initial and final mass) and using the following calculation, the percentage change in mass was obtained.

$$\frac{\text{Change in mass}}{\text{initial mass}} \times 100 = \text{percentage change in mass}$$

The entire experiment was repeated a further two times and the results averaged to improve their reliability.

The results are shown in the table below.

Concentration of sugar solution (%)	Initial mass (g)	Final mass (g)	Change in mass (g)	% change in mass	Average change in mass (%)
	2.0	2.3	0.3	15	
0	2.5	2.6	0.1	4	+9.1
	2.4	2.6	0.2	8.3	
	2.3	2.2	−0.1	−4.3	
5	2.2	2.1	−0.1	−4.5	−6.9
	2.5	2.2	−0.3	−12	
	2.2	1.9	−0.3	−13.6	
10	2.4	1.9	−0.5	−20.8	−15.5
	2.5	2.2	−0.3	−12	

Using the results, the following graph was drawn (fig 2.14):

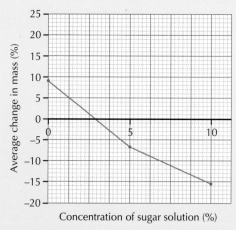

Fig 2.14 *Line graph of results from osmosis experiment*

From the results, it can be concluded that as the concentration of the sugar solution is increased, the average percentage change in mass decreases.

Complete the following tasks using the information from the osmosis experiment described.

1. The independent variable is the variable altered by the experimenter. Name the independent variable.

2. The dependent variable is the variable that changes as a result of altering the independent variable. Name the dependent variable.

3. Make a list of the variables that would need to be controlled in the experiment.

4. What was done in the experiment to ensure that the results obtained were reliable?

5. A similar experiment was repeated with different concentrations of sugar solutions. Again the experiment was repeated three times at each concentration.

The following results were obtained:

Concentration of sugar solution (%)	Initial mass (g)	Final mass (g)	Change in mass (g)	% change in mass	Average change in mass (%)
	2.0	2.1	0.1	5	
0.2	2.2	2.3	0.1		
	2.4	2.5	0.1	4.2	
	2.1	2.0	−0.1		
0.4	2.2	1.9	−0.2	−9.1	−6.0
	2.4	2.3	−0.1		
	2.3	2.1	−0.2	−8.7	
0.6	2.0	1.6	−0.4	−20	
	2.4	2.1	−0.3	−12.5	

(a) Copy the table and complete it using the calculation on page 36.

(b) Using your completed table, draw a graph to show the average change in mass (%) at each of the different sugar concentrations.

I can:

- Identify osmosis as a 'special case' of the diffusion of water.

- State that osmosis is the movement of water molecules from a higher water concentration to a lower water concentration through a selectively permeable membrane.

- Explain that animal cells burst when placed into solutions with a higher water concentration than the cell contains and shrink when placed into solutions with a lower water concentration than the cell contains.

- Explain that plant cells become turgid when placed into solutions with a higher water concentration than the cell contains and become plasmolysed when placed into solutions with a lower water concentration than the cell contains.

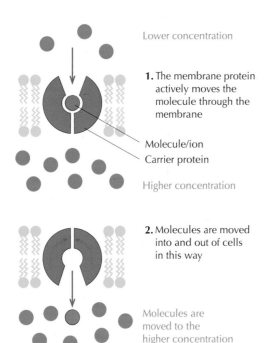

Lower concentration

1. The membrane protein actively moves the molecule through the membrane

Molecule/ion
Carrier protein

Higher concentration

2. Molecules are moved into and out of cells in this way

Molecules are moved to the higher concentration

Fig 2.15 *Active transport of molecules*

Active transport

Active transport is **the movement of molecules and ions from a region of lower concentration to a region of higher concentration**. This is the opposite of diffusion. Diffusion is a passive process so does not require any additional energy. **Active transport** has to work **against the concentration gradient to move molecules and ions**, which **requires additional energy**.

The membrane proteins act like pumps to move molecules and ions across the membrane. As this goes against the concentration gradient, the proteins use energy.

In nerve cells, sodium and potassium are actively transported across the cell membrane. The protein molecules in the membrane act as carriers to transport these substances. Iodine is also transported this way in seaweed.

Figure 2.15 shows molecules moving from a region of lower concentration to a region of higher concentration, with a protein acting as a pump.

Activities

Activity 1.2.6 Working individually

Write down full answers to the following questions:

1. Define the term 'active transport'.
2. Explain how active transport of molecules occurs across cell membranes.

Activity 1.2.7 Working in pairs

Draw a diagram to show the structure of the cell membrane. Include at least four membrane proteins with channels. Label the phospholipids and proteins. Split your diagram in half. On one side of the diagram show examples of passive transport; on the other show active transport. Your diagram should include examples of diffusion, osmosis and active transport. Use different shapes and colours to represent the molecules that would move across the cell membrane. Add as much detail as possible to show your understanding of these topics.

Activity 1.2.8 Working in pairs

Make a set of revision cards on this topic. Use coloured paper or different colours of pen. On one side of the card draw a picture or write the word that you want to learn. On the other side write the definition. For example, on one side write the word 'Diffusion'; on the other side write what the word means. Use the cards to test each other.

I can:

- State that active transport is the movement of molecules of a substance against a concentration gradient and requires energy.
- State that active transport is the movement of molecules of a substance from an area of lower concentration to an area of higher concentration.
- State that active transport requires membrane proteins to move substances against the concentration gradient.

3 DNA and the production of proteins

You should already know:

- Genes are located on chromosomes in the nucleus.
- Genes are made of DNA.
- DNA carries instructions to make proteins.
- Each individual's DNA is unique.
- Genes are passed on from parents to offspring.

Learning intentions

- Describe the structure of DNA.
- Name the four bases that make up the genetic code.
- Explain the relationship between DNA and proteins.
- Explain the relationship between the order of bases on DNA and the amino acids in a protein.
- Describe the role of mRNA in protein production.
- Name the basic units that proteins are made from and where protein synthesis takes place.

Chromosomes, genes and DNA

All of the genetic information is held on chromosomes found within the nucleus of almost all body cells. The chromosomes are made up of a chemical called deoxyribonucleic acid (DNA).

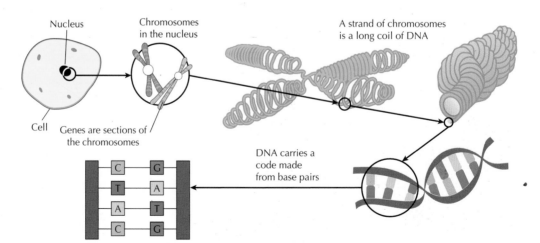

Fig 3.1 *How DNA is packaged into a cell. Where a black circle is shown, this indicates an area that has been magnified*

If we look at a chromosome in greater detail we can see the bands of different genes. A gene is a section of DNA which codes for a protein.

The structure of DNA

DNA is a molecule that consists of two strands connected together by bases. The DNA is described as a **double-stranded helix**, as it is found twisted and tightly packed together, rather like a coiled spring (fig 3.2).

The two strands are made up of chemicals called **bases**. There are four different bases found in DNA: adenine (**A**), thymine (**T**), guanine (**G**) and cytosine (**C**). The two strands of DNA join together when the bases join together. The bases always pair together in the same way; A joins with T, and G with C. This is known as **complementary base pairing** (fig 3.3). The bases make up the **genetic code**.

Fig 3.2 *Model of DNA*

🌳 Biology in context

Throughout history, scientists have been making discoveries about DNA. It was first isolated from the cell nucleus in 1869 by a biologist called Friedrich Miescher. Research continued by various scientists, and by the late 1940s it was accepted that DNA was the molecule which carried the genetic code. However, the structure of the molecule was then unknown.

It wasn't until 1953 that molecular biologists James Watson and Francis Crick produced a model showing the structure of DNA as a double helix (fig 3.4).

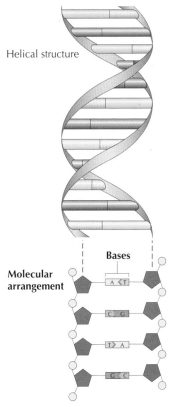

Helical structure

Molecular arrangement

Bases

A ⟨ T

C ⟩ G

T ⟩ A

G ⟨ C

Fig 3.3 *Complementary base pairing in DNA*

Fig 3.4 *Watson and Crick with a model showing the structure of DNA as a double helix*

(Continued)

They were later awarded a Nobel prize for the discovery, along with Maurice Wilkins. Watson and Crick based their ideas on earlier research carried out by Rosalind Franklin and Wilkins. Both Franklin and Wilkins carried out a lot of work on X-ray diffraction of DNA. It was an X-ray diffraction photograph of Franklin's that helped Watson and Crick to understand the helical structure of the molecule.

The base-pairing rule had first been explained in 1949 by a biochemist named Erwin Chargaff. He discovered that although different organisms had different quantities of DNA, the percentage of adenine was always equal to the percentage of thymine, as was the percentage of cytosine equal to the percentage of guanine.

Extracting DNA

DNA can be extracted from fruit and vegetables. The experiment below describes how DNA was extracted from a strawberry.

Method

1. A strawberry is added to a plastic bag and smashed to a pulp.

2. An extraction fluid is made by combining the following: 90 ml of water, 5 g of salt and 10 ml of detergent.

3. 30 ml of extraction fluid is added to the bag containing the strawberry and mixed well.

4. The bag is placed in a 60°C water bath for 15 minutes.

5. The bag is then placed into an ice bath for 5 minutes.

6. The mixture is filtered to remove any remaining solid strawberry.

7. The mixture is poured into a boiling tube until it is half full.

8. The boiling tube is tilted to the side, and ice-cold ethanol is slowly poured down the side of the tube.

Fig 3.5 *DNA extracted from a strawberry*

Result

The DNA is visible floating on top of the ethanol (fig 3.5).

GO! Activities

Activity 1.3.1 Working individually

Complete the following questions:

1. Describe the structure of DNA.
2. Name the four bases that make up the genetic code.
3. Name the complementary base pairs.
4. Give the definition of a gene.

Activity 1.3.2 Working in pairs

Firstly, use a mini whiteboard to 'thought-shower' what you know about DNA. Make a simple diagram showing DNA as a double-stranded molecule with four different bases.

Using this information, choose from **one** of the following tasks:

- Create a DNA model.

Use any material available to you. Suggested materials are: coloured plasticine/Play-Doh, coloured pipe cleaners/drinking straws, coloured paper or sweeties that could be linked together with cocktail sticks. The choice is yours!

- Produce a colourful DNA poster.

Whichever task you choose, you must show the DNA as a double-stranded molecule and the four different bases. You must also show complementary base pairing. You do not have to show the DNA as a twisted double helix, but you must show that it is double-stranded (fig 3.6).

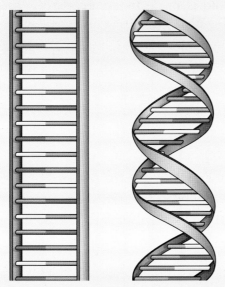

Fig 3.6 *DNA untwisted and twisted into a double helix*

Activity 1.3.3 Working individually or in pairs

Carry out some research into the history of DNA. Produce a timeline of the discovery of the molecule and a short biography of the main scientists involved.

I can:

- Describe the structure of DNA as being made up of a double-stranded helix held together by complementary base pairs.

- State that a gene is a section of DNA which codes for a protein.

- State that DNA contains four different bases: Adenine (A), Thymine (T), Guanine (G) and Cytosine (C) that make up the genetic code.

- Identify that base A pairs with base T, and base G pairs with base C; this is known as complementary base pairing.

- State that DNA carries the genetic information for making proteins which are unique to each individual.

Make the link – Biology

All humans have physical differences. We can account for these differences owing to our DNA coding for the different proteins made in our bodies. You will examine some of these differences in Area of study 2 when you study variation and inheritance (Chapter 10, page **136**).

DNA and proteins

The DNA carries the genetic information for making **proteins**. Proteins are made up of amino acid molecules joined together. A group of three bases on the DNA form a triplet.

Each triplet creates a code for a different amino acid (fig 3.7).

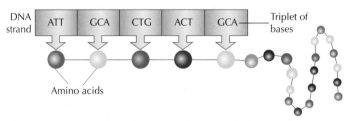

Fig 3.7 Diagram to show how DNA bases code for specific amino acids in protein

As there are four different bases (A, T, G and C), the triplet code allows for there to be a different code for each of the 20-plus different amino acids that make up all the proteins in nature.

The **type of protein made is determined by the sequence of bases** on the DNA. The sequence of bases on the DNA will provide the information for which amino acids are coded for. The **order of the amino acids then determines which protein is formed**. As each individual's DNA is different, the proteins coded for are unique to that individual.

Activity

Activity 1.3.4 Working individually

1. A section of DNA has 800 bases. Copy and complete the table below to show how many of each type of base there are and the percentage of each type of base.

Type of base	Number of bases	Percentage of bases
A	160	
T		
G	240	
C		30

2. Using the information in the table above, draw a bar chart to show the percentage of each type of base.

Hint

When drawing bar charts remember the following:

- labels on both axes
- a scale (with units) on the Y axis, with a value at the origin (usually a zero)
- bars should be of equal width and equally spaced
- always use a ruler!

Production of proteins

The information to produce proteins is held in the DNA in the nucleus. Proteins carry out a variety of jobs in the body and are essential for life. They are produced at ribosomes found in the cytoplasm of the cell.

As there is no way for the DNA to leave the nucleus, a copy of the code must be taken from the DNA in the nucleus.

Messenger RNA (**mRNA**) is a molecule which **carries a complementary copy of the genetic code from the DNA** held in the **nucleus** to a **ribosome** in the cytoplasm. From the copied code, the correct **amino acids are joined together**. The correct amino acid to match the code is identified using the complementary base-pairing rule. The amino acids form a long chain and eventually **a protein is formed** (fig 3.8).

The variety and functions of proteins are discussed in the next chapter.

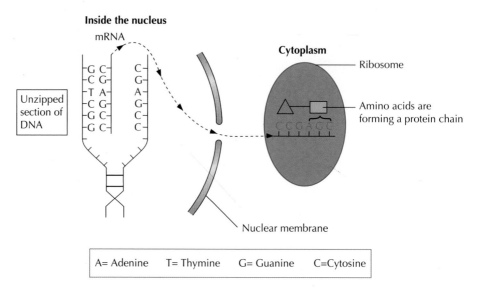

| A= Adenine | T= Thymine | G= Guanine | C=Cytosine |

Fig 3.8 *Protein chain being formed at a ribosome from mRNA code*

GO! **Activities**

Activity 1.3.5 Working individually

Write down full answers to the following questions:

1. Explain what the sequence of bases carried in the DNA determines.
2. Explain the role of the molecule mRNA.
3. Name the organelle where the protein is assembled from amino acids.

Activity 1.3.6 Working individually

Make a mini-poster to show a protein chain.

• From the four bases (A, T, G and C), make up a section of DNA consisting of 15 bases in any order of your choice. Show this on your poster.

• Use a different colour for each amino acid and show how a protein chain is formed.

I can:

• Explain that the sequence of bases determines the sequence of amino acids in a protein.

• State that messenger RNA (mRNA) is a molecule which carries a complementary copy of the genetic code from the DNA in the nucleus to a ribosome in the cytoplasm.

• State that at a ribosome, the mRNA is decoded, and the correct amino acids are assembled from the code.

• State that amino acids join together to form a protein.

4 Proteins

Variety and function of proteins

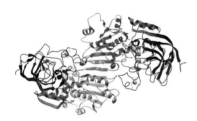

Fig 4.1 *Protein as a 3-dimensional shape*

Proteins are complex molecules that contain the chemical elements carbon, hydrogen, oxygen and nitrogen. The carbon, hydrogen, oxygen and nitrogen atoms are arranged to form different amino acids. There are many different types of proteins, and when they are examined at a molecular level they all have very different 3-dimensional shapes. Special computer programs are used to show the 3-dimensional shapes of proteins. An example of this is shown in figure 4.1. The protein looks very ribbon-like.

The variety of proteins is due to the **sequence of amino acids**, which is determined by the order of bases on the DNA. As there are 20 different amino acids, there many different ways in which they can be arranged.

Amino acids are joined together in a chain connected by peptide bonds. Long chains of amino acids form a polypeptide chain. The polypeptide chains twist around, forming many different shapes. Different combinations of amino acids can join together, making the possible combinations of proteins almost endless.

To illustrate this in a simple way, different shapes are often used to represent the different amino acids. Figure 4.2 shows how different amino acids are joined together to form proteins.

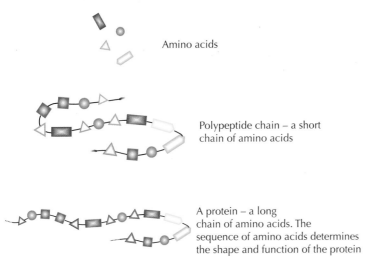

Amino acids

Polypeptide chain – a short chain of amino acids

A protein – a long chain of amino acids. The sequence of amino acids determines the shape and function of the protein

Fig 4.2 *Amino acids joined together to form a protein*

Proteins have many functions in the human body. One of these functions was explored in pages 26–27 of Chapter 2 when the structure of the cell membrane was examined. Proteins were shown to have an important **structural** function in cell membranes. Proteins are needed to support the membrane. Some have channels, and others act as carriers to allow the movement of substances into and out of cells (fig 4.3).

Some of the membrane proteins act as **receptors**. Receptors receive external signals and provide a binding site for molecules. This allows information to be passed to the inside of the cell.

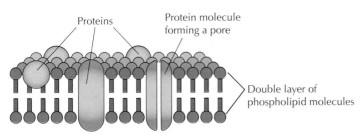

Proteins

Protein molecule forming a pore

Double layer of phospholipid molecules

Fig 4.3 *Structural function of proteins in the cell membrane*

Other proteins act as **enzymes**. These are important in cell metabolism – all of the chemical reactions that occur in cells. Enzymes are necessary to speed up the chemical reactions that are essential for life. The enzyme pepsin is important to break down proteins in the body. Its structure is shown in figure 4.4.

Some proteins also function as **hormones**. Hormones are chemical messengers that are important in the regulation of processes like growth and sugar balance in humans. The structure

Fig 4.4 *3-dimensional structure of the enzyme pepsin*

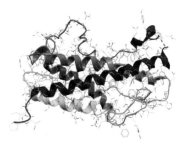

Fig 4.5 *3-dimensional structure of human growth hormone*

of human growth hormone is shown in figure 4.5. You can see how the structures of all proteins differ.

The body produces **antibodies**, which are made of protein, to fight against disease. Antibodies are produced in response to specific disease-causing microorganisms. A different antibody molecule is required for each microorganism. The structure of immunoglobulin (a type of antibody) is shown in figure 4.6.

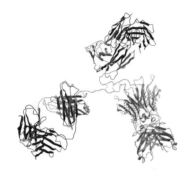

Fig 4.6 *3-dimensional structure of immunoglobulin*

GO! Activity

Activity 1.4.1 Working individually

1. Write down full answers to the following questions:
   (a) Explain how the variety of protein shapes and functions arises.
   (b) Describe the five main functions of proteins.

Protein function	Example
	Gives strength and support to cellular structures. E.g. Cell membranes
	Act as biological catalysts to speed up chemical reactions in cells. E.g. Amylase
	Carry chemical messages in the bloodstream. E.g. Insulin
	Produced by white blood cells to provide specific defence against certain bacteria and viruses. E.g. immunoglobulins
	Cell surface proteins that allow the cell to recognise specific substances. E.g. insulin receptor

2. Draw five different protein chains – one for each of the protein functions described above. Each chain should have at least six amino acids in it. Use a single line to join the amino acids to each other to represent the peptide bonds between them. Label your diagram as fully as possible.

I can:

- State that the variety of protein shapes and functions arises from the sequence of amino acids.
- Describe the functions of proteins as structural, receptors, enzymes, hormones and antibodies.

Enzymes and their functions

Enzymes are biological catalysts made by all living cells. A catalyst is a substance that speeds up a chemical reaction. An enzyme is classed as a **biological catalyst**, as it speeds up reactions that take place inside cells.

Although enzymes **speed up cellular reactions**, they themselves remain **unchanged** by the process and can be used over and over again.

The action of enzymes

The action of enzymes is demonstrated in the diagram below:

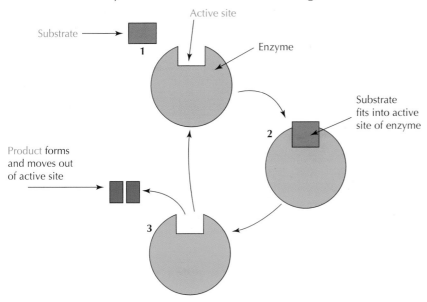

Fig 4.7 *Enzyme action*

1. The substance which a particular enzyme works on is called its **substrate**.

2. Enzyme molecules have a region of their surface known as the active site. The **active site** is the site of the chemical reaction on an enzyme. The substrate binds to the active site of an enzyme forming an enzyme-substrate complex. Only when this complex is formed will the reaction take place.

3. Enzyme action results in the formation of a **product(s)**. Once the reaction has taken place, the product(s) of the reaction are released, and the enzyme moves on to carry out the same reaction with more of the same substrate. This continues until there is no more of the substrate left.

Enzyme specificity

Each enzyme can catalyse only one reaction. This is due to their specific nature. Enzymes are specific due to the shape of their active site. The active site is complementary in shape to the specific substrate molecules and binds to them specifically. This is sometimes explained using the 'lock and key' hypothesis. On each enzyme there is an active site (the lock). Only one substrate (the key) can fit together with the active site.

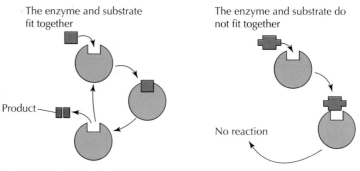

Fig 4.8 *Lock and key hypothesis*

Activity

Activity 1.4.2 Working individually

Write down full answers to the following questions:

1. State what enzymes are and where they can be found.
2. Describe the main function of an enzyme.
3. Define the terms 'active site', 'substrate' and 'product' in relation to enzymes.
4. Explain the relationship between the active site of an enzyme and its substrate.

I can:

- State that enzymes function as biological catalysts and are made by all living cells.
- State that enzymes speed up cellular reactions and are unchanged in the process.
- State that the active site is the site of the chemical reaction on an enzyme.
- State that a substrate is the substance that is acted upon by an enzyme.
- State that the shape of the active site of an enzyme molecule is complementary to its specific substrate(s).
- State that it is the formation of an enzyme–substrate complex that brings about a reaction.
- State that enzyme action results in product(s).

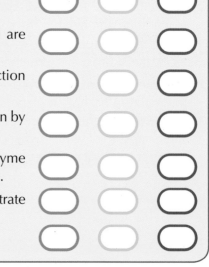

Enzyme reactions

Different enzymes are responsible for different reactions since the substrates they work on differ in shape. Enzymes can be involved in two different types of reaction: degradation (breakdown) and synthesis (build-up) reactions.

Degradation (breakdown) reactions

A degradation reaction involves large molecules being broken down into small molecules.

Examples of enzymes involved in degradation reactions are:

- Catalase, which is responsible for breaking down large molecules of hydrogen peroxide into smaller molecules of oxygen and water.

$$\text{Hydrogen Peroxide} \xrightarrow{\text{Catalase}} \text{Oxygen} + \text{Water}$$

- Amylase, which is responsible for breaking down large molecules of starch into smaller molecules of maltose.

$$\text{Starch} \xrightarrow{\text{Amylase}} \text{Maltose}$$

The action of amylase on a starch molecule is illustrated by the diagram below. The shape of the active site on the amylase molecule is complementary to the shape of the starch molecule. When combined amylase speeds up the breakdown of the large substrate molecule into smaller molecules, giving a new product, maltose. The reaction will continue until there is no more substrate molecule for the amylase to act upon.

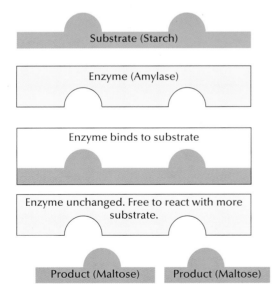

Substrate (Starch)

Enzyme (Amylase)

Enzyme binds to substrate

Enzyme unchanged. Free to react with more substrate.

Product (Maltose)   Product (Maltose)

Synthesis (build-up) reactions

A synthesis reaction involves small molecules being built up into a large molecule.

An example of an enzyme involved in a synthesis reaction is phosphorylase. Phosphorylase builds up small molecules of glucose-1-phosphate into larger molecules of starch.

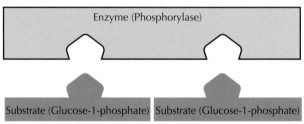

$$\text{Glucose-1-phosphate} \xrightarrow{\text{Phosphorylase}} \text{Starch}$$

The action of phosphorylase on glucose-1-phosphate molecules is illustrated by the diagram below. The shape of the active site on the phosphorylase molecule is complementary to the shape of the glucose-1-phosphate molecules. When combined phosphorylase speeds up the synthesis of the larger product molecule, starch, from the smaller substrate molecules. The reaction will continue until there is no more substrate molecule for the phosphorylase to act upon.

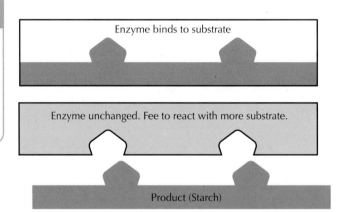

> ### 🔎 Hint
>
> Remember: Make sure you can give examples of enzymes involved in synthesis and degradation reactions, making reference to the enzyme involved, their substrate(s) and product(s).

Thinking about experiments ★1

Investigating enzyme action

An example of an enzyme found in all living cells is catalase. It speeds up the breakdown of hydrogen peroxide. Hydrogen peroxide is a poisonous substance, and its build-up inside the body would result in death.

Hydrogen peroxide can be broken down into oxygen and water – two harmless substances. To ensure that the reaction works fast enough to prevent the body from coming to harm, the enzyme catalase makes the reaction go faster.

The breakdown of hydrogen peroxide is demonstrated by adding a piece of liver tissue which contains catalase to some hydrogen peroxide. Bubbles of oxygen gas are produced (fig 4.9). By measuring the foam (oxygen bubbles) produced over a set period of time enzyme activity can be measured. This allows for the comparison of the same enzyme under different conditions.

The reaction is summarised in the following word equation:

$$\text{Hydrogen peroxide} \xrightarrow{\text{Catalase}} \text{Oxygen} + \text{Water}$$
$$(H_2O_2)$$

Substrate   Enzyme   Products

Fig 4.9 *The breakdown of hydrogen peroxide from liver tissue by the enzyme catalase*

🔍 Hint

To help you to remember that catalase speeds up the breakdown of hydrogen peroxide, remember **HP COW!**

$$\textbf{H}\text{ydrogen } \textbf{P}\text{eroxide} \xrightarrow{\text{Catalase}} \textbf{O}\text{xygen} + \textbf{W}\text{ater}$$
$$(H_2O_2)$$

GO! Activities

Activity 1.4.3 Working individually

Write down the answers to the following questions:

1. Name the two general types of reaction that involve enzymes.
2. Using a diagram, give an example of a reaction that involves the build-up of smaller molecules into a large molecule. Clearly label the enzyme involved and the specific substrate(s) and product(s).
3. Using a diagram, give an example of a reaction that involves the breakdown of a large molecule into smaller molecules. Clearly label the enzyme involved and the specific substrate(s) and product(s).

Activity 1.4.4 Working in pairs

Read the information below carefully and discuss the experiment.

A pupil carried out an experiment to find out which type of tissue contained the most catalase.

The experiment was set up as follows:

*   10 ml of hydrogen peroxide and two drops of detergent were added to four boiling tubes as shown.
*   Three discs of a different type of tissue were added to each boiling tube: carrot, apple, potato and boiled potato (fig 4.10). The shape, age and mass of all the tissues were kept the same.

(Continued)

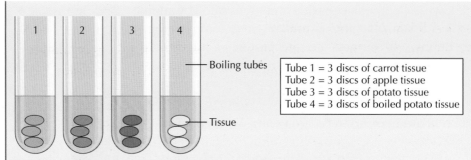

Tube 1 = 3 discs of carrot tissue
Tube 2 = 3 discs of apple tissue
Tube 3 = 3 discs of potato tissue
Tube 4 = 3 discs of boiled potato tissue

Fig 4.10 *Investigating catalase concentration in different types of tissue*

- The boiling tubes were left for 5 minutes, and then the height of foam produced in each tube was measured.
- The results are shown in the table below:

Type of tissue	Height of foam (mm)
Carrot	58
Apple	42
Potato	64
Boiled potato	0

1. Using the results in the table above, draw a bar chart to show the height of foam for each of the tissue types.

2. Read the information below and discuss it with your partner.

In any experiment a scientist must ensure the following:

- **Validity**: This is to do with the 'fairness' of the experimental procedure. In any experiment, only one factor (variable) should be varied, and all other factors (variables) that could affect the results should be kept the same.
- **Reliability**: To ensure reliability of an experiment and the results, the experiment is repeated, and several readings or samples are taken.
- **Controls**: A control is identical to the original experiment in every way except that the factor that causes the change is removed. This is then replaced by something that does not bring about change.

Write down full answers to the following questions:

 (i) What did the pupil do to ensure the validity of the experiment?
 (ii) Which variables were kept the same?
 (iii) What did/could the pupil do to ensure the reliability of the procedure?
 (iv) Was a control used in the experiment? What would it tell us?
 (v) What conclusions can be drawn from this experiment?

Activity 1.4.5 Working individually

Read the information below carefully, and then complete the tasks that follow.

A pupil carried out an experiment to investigate the specificity of enzymes.

The experiment was set up as follows:

- 10 ml of hydrogen peroxide and two drops of detergent were added to four boiling tubes (fig 4.11). The same bottle of hydrogen peroxide and detergent was used each time, and all boiling tubes were of the same diameter.

- 1 ml of a different enzyme was added to each boiling tube: amylase (tube 1), catalase (tube 2), lipase (tube 3), pepsin (tube 4).

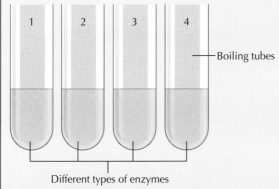

Different types of enzymes

Fig 4.11 *Investigating enzyme specificity*

- The boiling tubes were left for 5 minutes, and then the height of foam produced in each tube was measured.

- The results are shown in the table below:

Enzyme	Height of foam (mm)
Amylase	0
Catalase	50
Lipase	0
Pepsin	0

Write down full answers to the following questions:

1. What conclusions can be drawn from the results obtained?
2. What has been done to ensure the validity of the experiment?
3. Which variables were kept the same?
4. What could be/is done to ensure the reliability of the procedure?
5. Was a control used in the experiment? If not, what would a suitable control be? What would it tell us?

I can:

- State that enzymes can be involved in degradation and synthesis reactions.
- Give examples of synthesis and degradation reactions to relate the enzymes involved to their specific substrate(s) and product(s).
- Use diagrams to illustrate the stages in degradation and synthesis reactions, e.g.

$$\text{substrate} \xrightarrow{\text{enzyme}} \text{product}$$

Conditions affecting enzyme activity

Each enzyme is most active under **optimum** conditions. At its optimum an enzyme is at its most active – the rate of reaction cannot go any faster. Although the optimum conditions vary for each enzyme, enzymes and other proteins are affected by **temperature** and **pH**.

If the shape of an enzyme's active site is altered it is said to be **denatured** and will no longer fit together with its substrate.

If an enzyme–substrate complex cannot be formed, this will affect the rate of the reaction. Without an enzyme to speed up the reaction it will take place at a much slower rate.

The experiment shown in figure 4.12 illustrates the effect of different temperatures on the action of catalase.

Liver tissue (source of catalase) is added to hydrogen peroxide that ranges in temperature from 5°C to 60°C. Catalase speeds up the breakdown of hydrogen peroxide, releasing oxygen bubbles.

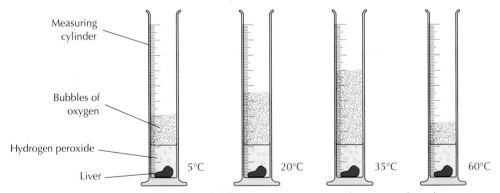

Fig 4.12 *The effect of different temperatures on the action of catalase*

At 5°C and 60°C there is low oxygen production, as the conditions are not optimal for catalase activity. At 35°C most oxygen bubbles are produced.

The optimum temperature and pH of each enzyme are normally linked to the organism in which it is found. Most enzymes found in the human body are most active at 37°C (body temperature). Plant enzymes on the other hand are often most active at around 20°C.

At low temperatures, the rate of a reaction is decreased as enzyme molecules have less energy to move around. As the temperature is increased, the rate of enzyme activity increases until it reaches the optimum. This is demonstrated in figure 4.13, where a graph showing the typical effects of temperature on enzyme activity in the human body is illustrated.

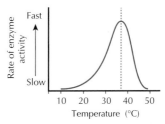

Above the optimum, as temperature continues to increase the rate of enzyme activity decreases. The increased temperature causes the shape of the active site to change. This makes it difficult for the substrate to bind with the enzyme. If the temperature continues to increase (over 50°C), the shape of the active site is permanently changed and the enzyme is **denatured**.

Fig 4.13 *Enzyme activity and temperature*

Different enzymes have different optimum pH levels. The enzyme pepsin works best around pH 2. It is found in the stomach where conditions are very acidic. Catalase found in the liver works best at pH 9 (fig 4.14).

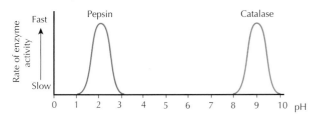

Fig 4.14 *Enzyme activity and pH*

Graphs of enzyme activity with increasing temperature or pH all have a similar shape. Enzyme activity increases until the optimum and then decreases (fig 4.15).

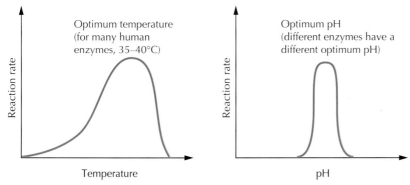

Fig 4.15 *Graphs of optimum temperature and pH*

🌳 Biology in context

You may or may not have been aware of it, but at some point you will most likely have observed the effects of temperature on proteins.

If you have boiled, fried or poached an egg, you will have observed the effects of temperature on a protein.

The white of an egg (egg albumen) is made of protein. In a raw egg (fig 4.16a) the albumen is a clear, runny liquid. As the albumen is heated, its consistency changes to become a white solid (fig 4.16b). Once heated, the albumen is permanently altered and cannot be changed back.

Fig 4.16a *Raw egg*   **Fig 4.16b** *Fried egg*

Enzymes are also made of protein and so are irreversibly altered once heated. When heated to over 50°C the shape of the active site is permanently changed. This causes the enzyme to become denatured.

It is important to remember that although enzymes are made of protein, not all proteins are enzymes.

GO! Activities

Activity 1.4.6 Working individually

Read the information below carefully, and then complete the tasks that follow.

An experiment was carried out to investigate the effect of pH on enzyme activity.

Hydrogen peroxide is used as the substrate and catalase as the enzyme. Catalase speeds up the breakdown of hydrogen peroxide and is most active at its optimum pH. The breakdown of hydrogen peroxide results in the formation of oxygen, so the height of foam produced is used as a measure of enzyme activity.

The experiment was set up as follows:

- 10 ml of hydrogen peroxide and two drops of detergent were added to six boiling tubes.
- 2 ml of six different pH solutions were mixed with the hydrogen peroxide ranging from pH 6 to 11.

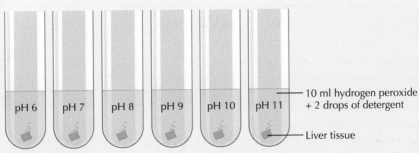

pH 6   pH 7   pH 8   pH 9   pH 10   pH 11

10 ml hydrogen peroxide + 2 drops of detergent

Liver tissue

Fig 4.17 *Effect of pH on the breakdown of hydrogen peroxide*

- A 2-gram piece of liver tissue (a source of catalase) was then added to each boiling tube (fig 4.17).
- The boiling tubes were left for 5 minutes, and then the height of foam produced in each tube was measured.
- The results are shown in the table below.

pH	Height of foam (mm)
6	0
7	15
8	30
9	20
10	15
11	0

Complete the following tasks:

1. Using the information given in the table above, draw a line graph to show the height of foam produced at each pH.
2. Using the information given in the table state the pH at which the enzyme reaches its optimum.

Activity 1.4.7 Working individually

Produce an information leaflet about enzymes. You must include the following information:

- what enzymes are and where they are found
- the main function of an enzyme
- define the terms 'active site' and 'substrate'
- explain the relationship between the active site of an enzyme and its substrate
- a diagram to illustrate the 'lock and key' hypothesis
- give the meaning of the term 'optimum'
- name the factors that affect enzymes, and describe their effect
- define the term 'denaturation', and explain when it happens

(Continued)

Activity 1.4.8 Working in pairs

Carry out some research into enzymes and their optimum conditions. Find out about as many enzymes as you can. Construct a table of the information listing the name of the enzyme, its substrate, optimum temperature and optimum pH.

🔍 Hint

You may wish to use information you have gathered from researching enzymes to plan and carry out an experiment, to gather data for your assignment.

💥 Make the link – Biology

In Area of study 2 (Chapter 13) you will find out about the absorption of materials into the bloodstream. Enzymes play an important role in the digestion of food, helping to break down large insoluble molecules into small soluble molecules.

🌳 Biology in context

Enzymes have many industrial uses.

Many products and processes we use on a daily basis depend upon the action of enzymes. The production of cheese, baking of bread, brewing of beer and wine, and breakdown of waste are just some examples.

Enzymes are also used in biological washing powders to digest protein, starch and fat stains. Wash cycles normally use temperatures of 40–90°C. As enzymes work well at lower temperatures clothes can be washed on cooler wash cycles (30–40°C) and be cleaned just as well as or better than with non-biological washing powders. This is better for the environment and helps to save energy.

Fig 4.18 *Enzymes are used in biological washing powders*

I can:

- State that each enzyme is most active in its optimum conditions. ⬭ ⬭ ⬭

- Explain that enzymes can be denatured, resulting in a change in their shape, which will affect the rate of reaction. ⬭ ⬭ ⬭

5 Genetic engineering

You should already know:

- Cells can have therapeutic uses.
- There are moral and ethical issues associated with some controversial biological procedures.
- Genetic engineering can be used to produce insulin and other proteins.

Learning intentions

- Name the process used to transfer genetic information from one cell to another.
- Describe the process of genetic engineering.
- Explain the stages involved in the process of genetic engineering.
- Explain the use of enzymes in the process of genetic engineering.

Genetic engineering

Genetic information can be transferred from one cell to another by genetic engineering. It allows for the transfer of genes from one organism to another. Scientists have used this process to genetically modify food and for the production of medicines.

Some examples include (figs 5.1–5.6):

Fig 5.1 *Golden rice*

Fig 5.2 *Less toxic rapeseed oil*

Fig 5.3 *Bird resistance to bird flu*

Fig 5.4 *Tomatoes with a longer shelf-life*

Fig 5.5 *Blight-resistant potatoes*

Fig 5.6 *Production of medicines, e.g. insulin and human growth hormone*

Stages in genetic engineering

The steps involved in the process of genetically engineering **bacteria** so that they contain a **human** gene are shown below.

1. Identify the section of DNA that contains the required gene from the source chromosome.

2. Extract the required gene.

3. Extract the plasmid from the bacterial cell.

4. Insert the required gene into a bacterial plasmid.

5. Insert the plasmid into a host bacterial cell to produce a genetically modified (GM) organism.

Steps 2, 3 and 4 require the use of enzymes. In step 2 enzymes are used to remove the required gene from the source chromosome. In step 3 enzymes are required to cut open the plasmid, following its extraction from the bacterial cell. In step 4 enzymes are used to seal the required gene into the plasmid.

🔍 Hint

Questions on this topic will test your knowledge of the steps involved in genetic engineering. Make sure that you know the correct order and that you can identify when enzymes are used.

The use of genetic engineering in the production of insulin

The process of genetic engineering is used to produce insulin for humans. Insulin is normally made in the pancreas. It is a hormone (made of protein) that controls the levels of glucose (sugar) in the blood. Some people are unable to make insulin; this is called type 1 diabetes. Diabetics need to inject insulin to control their blood sugar level (fig 5.7).

Before genetic engineering, insulin was extracted from pigs. Not only was it a slow and expensive process, but pig insulin is not exactly the same as human insulin. Some people suffered from side effects, and others were uncomfortable with the use of animals.

Fig 5.7 *Diabetic injecting the hormone insulin*

Figure 5.8 shows the stages involved in the production of insulin using genetic engineering.

- The chromosome containing the insulin gene is extracted from a human cell, and the insulin gene is removed using enzymes.

- A plasmid is extracted from a bacterial cell and cut open using enzymes.

- The insulin gene is inserted and sealed into the plasmid using enzymes. The plasmid now contains the genetic information to make insulin, as well as the normal bacterial information.

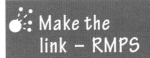

💥 Make the link – RMPS

The technology used in genetic engineering raises a number of ethical issues.

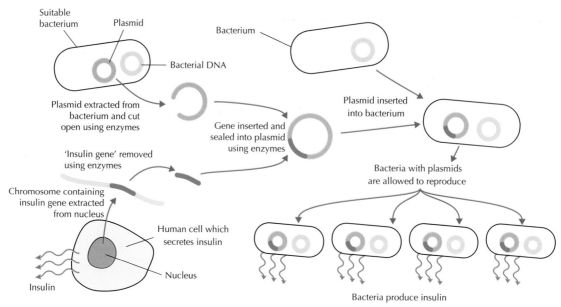

Fig 5.8 *Stages of genetic engineering*

- The plasmid is inserted back into the bacterial cell, which is allowed to reproduce.

- The genetically modified bacteria are grown in industrial fermenters to produce large volumes of insulin.

- The insulin can then be extracted and purified from the bacterial cells and used to treat diabetics.

GO! Activities

Activity 1.5.1 Working individually

Construct a table to show the list of genetically engineered food and medicines mentioned in the text. For each example explain the benefit to humans of the genetic modification.

Activity 1.5.2 Working individually

Investigate the arguments for and against genetic engineering. Produce an essay that clearly states both points of view. After careful consideration of both sides of the argument state your own personal viewpoint on the topic.

Activity 1.5.3 Working in pairs

Produce a poster to show the stages involved in genetically altering an organism. Include diagrams.

Activity 1.5.4 Working in pairs

Use the internet to research some of the more extreme examples of genetic engineering, e.g. Enviropig, venomous cabbage, glow-in-the-dark cats and medicinal eggs.

Produce a fact file giving four examples of extreme genetic engineering.

Include the following information:

- The name of the organism
- An explanation of how the organism has been genetically modified
- What the benefits are of the modification.

I can:

- Name genetic engineering as the process used to transfer genetic information from one cell to another.

- Describe the process of genetic engineering as the transfer of genetic information from one cell to another.

- Explain the process of genetic engineering by giving details of the stages involved: (1) identify the section of DNA that contains the required gene from the source chromosome; (2) extract the required gene; (3) extract the plasmid from the bacterial cell; (4) insert the required gene into a bacterial plasmid; (5) insert the plasmid into a host bacterial cell to produce a genetically modified (GM) organism.

- Explain that enzymes are used in the process of genetic engineering to remove the required gene from the source chromosome, to cut open the bacterial plasmid and to seal the required gene into the plasmid.

6 Respiration

Cellular respiration

Energy is required by all cells in the body for even the simplest of things. We need energy for movement, for keeping warm and even to keep our bodies working while we sleep.

Respiration is the chemical release of energy from the food that we eat. To ensure survival, all cells need to be able to carry out respiration.

To find out how much energy is present in different types of food, a simple experiment can be carried out. A food sample is burned under a set volume of water and the temperature increase measured. The energy contained in different food types can then be compared to determine which contains the most energy (fig 6.1).

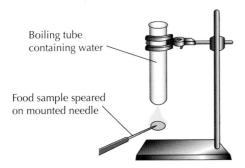

Boiling tube containing water

Food sample speared on mounted needle

Fig 6.1 *Measuring the energy content of food*

The chemical energy stored in **glucose** is released from cells through a series of **enzyme-controlled** reactions.

Respiration carried out in the presence of oxygen is called aerobic respiration. This can be summarised in the following equation:

Glucose + Oxygen → Carbon dioxide + Water + Energy (ATP)

The raw materials consist of glucose (from the food that we eat) and oxygen (from the air). These are combined to produce carbon dioxide (which we breathe out), water (which is also released in the breath) and energy in the form of ATP.

● Make the link – Biology

Respiration should not be confused with breathing. Breathing is a mechanism that allows the exchange of gases. See Area of study 2 page **194** for details of gas exchange in the lungs.

● Make the link – Biology

Respiration is controlled by enzymes. Enzymes are affected by temperature (Chapter 4).

● Make the link – Biology

Look back to Chapter 1 and find the link between respiration and mitochondria.

GO! Activity

Activity 1.6.1 Working individually

Write down full answers to the following questions:

1. Define 'respiration'.
2. Name the source of stored energy in cells.
3. The process of respiration involves a series of reactions. Describe how these reactions are controlled.

I can:

- Explain that the chemical energy stored in glucose must be released by all cells through a series of enzyme-controlled reactions called respiration.

The release and use of energy

The energy released from the breakdown of glucose is used to generate ATP.

ATP (adenosine triphosphate) is a high-energy molecule used to transfer chemical energy in living cells.

1 molecule of adenosine

Adenosine —— P_i —— P_i —— P_i

3 inorganic phosphate groups

Fig 6.2a *Chemical structure of ATP*

ATP is made up of one adenosine molecule attached to three inorganic phosphate molecules (P_i) (fig 6.2a).

The chemical energy contained in ATP can be transferred when required by cells.

Figure 6.2b summarises the link between energy and ATP.

The energy transferred by ATP can be used by cells for cellular activities including:

- muscle cell contraction
- cell division
- protein synthesis
- transmission of nerve impulses

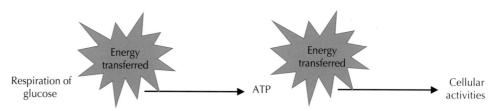

Respiration of glucose → Energy transferred → ATP → Energy transferred → Cellular activities

Fig 6.2b *The role of ATP in the transfer of energy*

GO! Activity

Activity 1.6.2 Working individually

1. Draw a simple diagram to show the link between ATP, the energy from respiration and the energy used for cellular activities.

2. Draw a spider diagram to show examples of cellular activities that require ATP.

I can:

- State that the energy released from the breakdown of glucose is used to generate ATP.

- State that the energy transferred by ATP can be used for cellular activities, including muscle cell contraction, cell division, protein synthesis and transmission of nerve impulses.

Aerobic respiration

Respiration involves a series of enzyme-controlled reactions to break down glucose. Complete breakdown of glucose requires aerobic conditions: respiration in the presence of oxygen.

The basic word equation for aerobic respiration can be summarised as:

Glucose + oxygen → carbon dioxide + water + energy (ATP)

This takes place in stages that are controlled by enzymes.

Stage 1: The breakdown of glucose

The first step in releasing the energy from glucose is to break it down to a substance called pyruvate. Each glucose molecule is broken down to two molecules of pyruvate. This takes place in the **cytoplasm** of the cell and does not require oxygen. Two ATP molecules are formed as a result of breaking glucose down to pyruvate (fig 6.3). Further breakdown of pyruvate depends on the presence or absence of oxygen.

Stage 1:

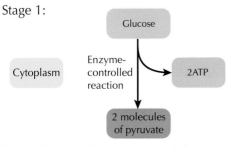

Fig 6.3 *Stage 1: The breakdown of glucose*

Stage 2: The breakdown of pyruvate

The second stage involves the breakdown of pyruvate to produce carbon dioxide and water. This stage can only take place if oxygen is present and takes place in the **mitochondria**. Each pyruvate molecule is broken down to release enough energy to yield a large number of ATP molecules (fig 6.4).

Stage 2:

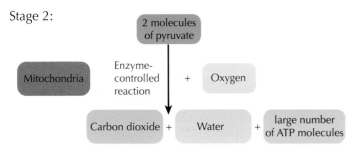

Fig 6.4 *Stage 2: The breakdown of pyruvate*

> **🔍 Hint**
>
> You may read or hear people talk about glycolysis. This is another name for the the first stage of respiration. You do not need to learn this term for National 5, but may wish to be familiar with it if you aim to progress to Higher level.

> **🔍 Hint**
>
> At this level you are expected to know that the breakdown of glucose to pyruvate yields two molecules of ATP. It is sufficient to understand that aerobic respiration results in the yield of a large number of ATP molecules.

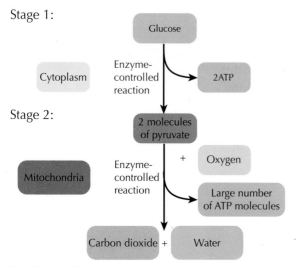

Stage 1:

Stage 2:

Fig 6.5 *Aerobic respiration*

Complete aerobic respiration

If we combine both stages we can see that the complete aerobic breakdown of one glucose molecule results in the yield of a large number of ATP molecules.

A summary of the whole process is shown in figure 6.5.

GO! Activities

Activity 1.6.3 Working individually

Write down full answers to the following questions:

1. Name the molecule that glucose is broken down to in the first stage of respiration.
2. State the number of ATP molecules yielded by the end of the first stage of respiration.
3. Name the final products that result from the complete breakdown of glucose.
4. Give the summary word equation for aerobic respiration.

Activity 1.6.4 Working in pairs

Using mini whiteboards or paper, take turns to practise drawing a diagram to show the complete breakdown of glucose in the presence of oxygen. Check each other's work.

I can:

- State that the summary word equation for respiration with oxygen is:

  Glucose + Oxygen → Carbon dioxide + Water + Energy

- State that glucose is broken down to two molecules of pyruvate, releasing enough energy to yield two molecules of ATP (further breakdown of pyruvate depends on the presence or absence of oxygen).

- State that in the presence of oxygen, aerobic respiration takes place and each pyruvate molecule is broken down to carbon dioxide and water, releasing enough energy to yield a large number of ATP molecules.

Measuring the rate of respiration in animals

The rate of respiration can be measured using a respirometer. In the simple respirometer shown in figure 6.6, a woodlouse is placed inside a sealed tube.

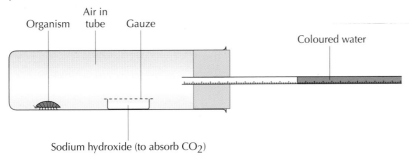

Fig 6.6 *A simple respirometer*

The only air available to the organism is inside the tube. A solution of sodium hydroxide is used to absorb any carbon dioxide produced by the woodlouse. This way, only the oxygen used up by the organism is measured. As the organism uses up oxygen, the volume of gas inside the tube decreases. The coloured water moves down the tube to fill the space of the oxygen that has been used up by the organism. The distance that the liquid moves in a set time period is used to calculate the rate of respiration.

Respiration in germinating seeds

Germination is the growth of a seed. To enable seeds to grow they must carry out respiration. This allows the food stored within the seed to be broken down to release energy. This energy is then used for the growth of a seed into a plant.

To prove that seeds carry out respiration the following experiment is set up. The chemical limewater is included in the experiment, as it changes from clear to cloudy in the presence of carbon dioxide. A thermometer is used to record the temperature as heat energy may be released from cells during respiration (fig 6.7).

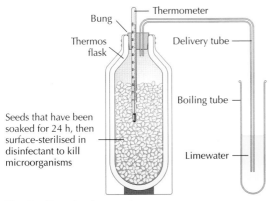

Fig 6.7 *Germinating seeds*

At the start of the experiment the temperature inside the flask is 20°C, and the limewater is clear. After 3 days the temperature has risen by 3°C, and the limewater is cloudy.

GO! Activity

Activity 1.6.5 Working individually

Using the information given about the experiment described above complete the following tasks.

1. **(a)** Construct a table to present the results of the experiment.

   (b) Explain how the results support the theory that the seeds are carrying out respiration.

   (c) Design a control experiment to show that it is the seeds that have caused the described changes to the limewater and temperature.

2. Investigate your own rate of respiration by measuring your breathing rate before and after exercise.

   • While at rest, count the number of breaths taken in 1 minute.

   • Do 3 minutes of exercise, e.g. jogging on the spot, star jumps or stepping up and down.

   • Immediately count the number of breaths taken in 1 minute.

   • Draw a bar chart of your results.

   Answer the following questions:

   1. What could you do to make the results more reliable?
   2. Would another person get exactly the same results?

Fermentation in animal cells

In the absence of oxygen, respiration cannot take place as it normally would. Therefore, glucose cannot be completely broken down to carbon dioxide and water as it would when oxygen is present.

The **fermentation pathway** is followed and takes place in the cytoplasm without oxygen.

The first stage of respiration takes place as normal, as it does not require oxygen. Following this fermentation pathway results in the break down of each glucose molecule to pyruvate and yields only the initial two molecules of ATP. In the absence of oxygen, pyruvate is then converted to **lactate**, as shown in figure 6.8. The whole process of fermentation is completed in the cytoplasm.

This pathway is followed by animal cells when their demand for oxygen cannot be met by the body. This happens frequently during strenuous exercise.

When a person first starts to exercise the body will respire aerobically. As the intensity of the exercise increases, the breathing rate increases to a point where gas exchange cannot take place any faster. Energy is still required by the body cells, so respiration proceeds without oxygen. This allows energy to be released, but as a result lactate builds up in the muscles.

Eventually, the muscles tire, and exercise has to stop. Following a period of recovery where oxygen is available, the lactate is converted back to pyruvate.

Fermentation in animal cells can be summarised in the following word equation:

Glucose → lactate + energy (2 ATP)

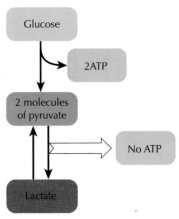

Fig 6.8 *Fermentation in animal cells*

🔍 Hint

Think of lactate as an oxygen 'debt'. Lactate is allowed to build up in the muscles when there is a lack of oxygen. As soon as oxygen is available it is 'repaid', and the lactate is removed.

GO! Activity

Activity 1.6.6 Working individually
Write down full answers to the following questions:

1. Describe the process of fermentation in animal cells.
2. State the number of ATP molecules produced during fermentation in animal cells.
3. Name the final product that results from fermentation in animal cells.
4. Give the summary word equation for fermentation in animal cells.

⚫ Make the link – PE, Health and Wellbeing

The demand for energy increases during exercise. This increases the demand for oxygen. This is why breathing rate and depth increase during exercise.

I can:

- State that in the absence of oxygen, the fermentation pathway takes place. ⬭ ⬭ ⬭

- State that the breakdown of each glucose molecule via the fermentation pathway yields only the initial two molecules of ATP. ⬭ ⬭ ⬭

- State that during fermentation in animal cells the pyruvate molecules are converted to lactate. ⬭ ⬭ ⬭

- State that the summary word equation for fermentation in animal cells is:

  Glucose → Lactate + Energy (2 ATP) ⬭ ⬭ ⬭

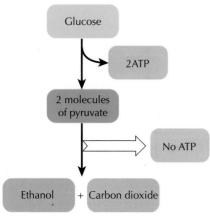

Fig 6.9 *Fermentation in plant and yeast cells*

Fermentation in plant and yeast cells

In plant and yeast cells a similar situation arises. The fermentation pathway is followed. The first stage of respiration takes place in the cytoplasm, without oxygen, and glucose is broken down to pyruvate. Again, only the initial two molecules of ATP are produced as a result of this process.

The difference between animal cells and plant and yeast cells occurs in the next stage. In plant and yeast cells, pyruvate is converted to **ethanol** (alcohol) and **carbon dioxide** as shown in figure 6.9.

As the carbon dioxide produced is released the reaction cannot be reversed.

Fermentation in plant and yeast cells can be summarised by the following word equation:

Glucose → ethanol + carbon dioxide + energy (2 ATP)

GO! Activity

Activity 1.6.7 Working individually

Write down full answers to the following questions:

1. Describe the process of fermentation in plant and yeast cells.
2. State the number of ATP molecules produced during fermentation in plant and yeast cells.
3. Name the final products that result from fermentation in plant and yeast cells.
4. Give the summary word equation for fermentation in plant and yeast cells.

I can:

- State that the breakdown of each glucose molecule via the fermentation pathway yields only the initial two molecules of ATP.
- State that during fermentation in plant and yeast cells the pyruvate molecules are converted to carbon dioxide and ethanol.
- State that the summary word equation for fermentation in plant and yeast cells is:

    Glucose → Carbon dioxide + Ethanol + Energy (2 ATP)

Biology in context

The process of fermentation is used commercially in industry.

Yeast cells are used in the brewing industry to produce alcohol in beer and wine.

Yeast is also used in bread making, as the production of carbon dioxide makes the dough rise (fig 6.10).

Fig 6.10 *Products of fermentation*

Location of aerobic respiration and fermentation

Respiration begins in the **cytoplasm**. The process of fermentation is completed in the cytoplasm, whereas aerobic respiration is completed in the **mitochondria**. Mitochondria are sausage-shaped organelles found in the cytoplasm of a cell.

The detail of a mitochondrion as revealed using a high-powered microscope is shown below (fig 6.11).

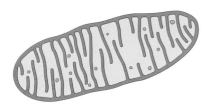

Fig 6.11 *The detail of a mitochondrion*

Make the link – Biology

Look back to Chapter 1 to see where and in which cell types we find mitochondria.

The higher the energy requirement of a cell, the greater the number of mitochondria present in that cell. Muscle, sperm and nerve cells all have many mitochondria present in the cytoplasm of their cells as they require a lot of energy.

GO! Activities

Activity 1.6.8 Working individually

1. Copy and complete the table below to compare aerobic respiration and fermentation.

	Aerobic respiration	Fermentation in	
		Animals	Plants and yeast
Product(s)			
Number of ATP molecules produced (per molecule of glucose)			

2. Choose a cell type that contains many mitochondria. Produce an information leaflet on the uses of energy in your chosen cell.

    You should include:

    *   a drawing of the cell showing many mitochondria present
    *   an explanation of how energy released from respiration is used to build up and break down ATP.

Activity 1.6.9 Working in pairs

Produce a poster to compare aerobic respiration and fermentation. Divide the poster into three sections: one for aerobic respiration, one for fermentation in animal cells, and the final section for fermentation in plant and yeast cells.

You should include:

*   as many differences between the processes as possible
*   the names of the parts of the cell where each process takes place
*   diagrams where possible
*   word equations for all processes.

I can:

- State that aerobic respiration begins in the cytoplasm and is completed in the mitochondria.
- State that the whole process of fermentation is completed in the cytoplasm.
- State that the higher the energy requirements of a cell the greater the number of mitochondria present in that cell.

Cell biology – Review questions

Section A

1. Which two structural features are common to both plant and bacterial cells?

    A Cell wall and ribosomes

    B Cell wall and nucleus

    C Chloroplasts and ribosomes

    D Sap vacuole and nucleus

2. Which of the following are two components of the cell membrane?

    A Phospholipids and carbohydrates

    B Proteins and cellulose

    C Carbohydrates and proteins

    D Proteins and phospholipids

3. Four cylinders of potato tissue were weighed and each was placed into a sugar solution of different concentration. The cylinders were re-weighed after 1 hour.

    The results are shown in the following table.

	Mass of potato cylinder (g)	
Sugar solution	**Initial mass**	**Final mass**
A	10.1	12.5
B	9.4	11.1
C	9.6	6.9
D	9.8	6.7

    In which sugar solution would most potato cells be expected to be plasmolysed?

4. Passive transport occurs

    A against a concentration gradient and does not require additional energy

    B down a concentration gradient and does not require additional energy

    C down a concentration gradient and requires additional energy

    D against a concentration gradient and requires additional energy

5. The diagram below shows a section of unwound DNA.

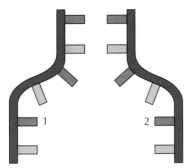

If number 1 represents the base G, then number 2 must represent the base

A   G

B   A

C   T

D   C

6. The DNA of a gene carries information that determines the structure of

A   bases

B   fats

C   proteins

D   carbohydrates

7. A fragment of DNA was found to have 120 cytosine bases and 60 thymine bases. What is the total number of bases in this fragment?

A   60

B   90

C   180

D   360

8. The active site of an enzyme is complementary to

A   one type of product molecule

B   all types of product molecules

C   one type of substrate molecule

D   all types of substrate molecules

9. The steps involved in the process of genetic engineering are shown below:

1   Extract the plasmid from the bacterial cell.

2   Extract the required gene.

3   Insert the plasmid into a host bacterial cell to produce a genetically modified (GM) organism.

4   Insert the required gene into a bacterial plasmid.

5   Identify the section of DNA that contains the required gene from the source chromosome.

The correct order for these steps is

A   3, 5, 2, 4, 1

B   2, 5, 1, 4, 3

C   5, 2, 1, 4, 3

D   1, 4, 3, 2, 5

10. Which line in the table identifies correctly the location and products of fermentation in plant cells?

	Location	Products
A	Mitochondria	Water + carbon dioxide
B	Cytoplasm	Ethanol + carbon dioxide
C	Mitochondria	Ethanol + carbon dioxide
D	Cytoplasm	Water + carbon dioxide

Section B

1. The diagram below shows a yeast cell.

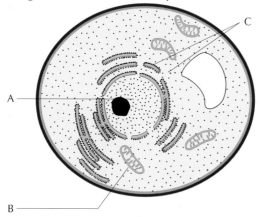

(a) Copy and complete the table below to give the name and function of the parts labelled A, B and C.

Letter	Structure	Function
A	Nucleus	
B	Mitochondrion	
C		Site of protein synthesis

2

(b) Name the structure that is present in green plant cells but is absent from yeast cells.

_____ 1

(c) A number of different cell types were viewed under a microscope. The cells were measured and the average cell sizes were calculated. The results are shown in the table below.

Type of cell	Average length of cell (μm)
Animal	23
Plant	47
Bacterial	2
Fungal	6

On a separate piece of paper, copy and complete the vertical axis and draw a bar chart to show the average lengths of the cells shown in the table.

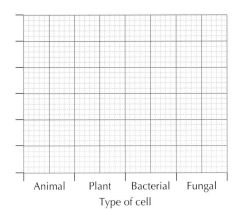

Animal   Plant   Bacterial   Fungal
Type of cell

2

2. The diagram below represents the transfer of oxygen from the atmosphere to working muscle cells.

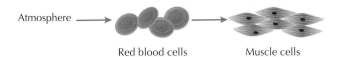

Atmosphere →   Red blood cells   Muscle cells

(a) From the diagram, state where the highest and the lowest concentrations of oxygen would be expected.

Highest concentration _____

Lowest concentration _____ 1

(b) Name the process by which oxygen enters cells.

_____ **1**

3. Describe the osmotic effect of transferring:
   Animal cells from a weak salt solution to a solution of pure water.
   or
   Plant cells from a weak sugar solution to a strong sugar solution.

   _____

   _____

   _____

   _____

   _____ **3**

4. The diagram below shows how a protein chain is assembled from the DNA code.

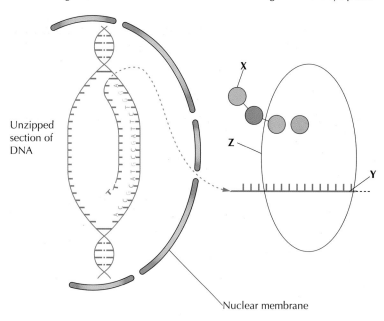

Stage 1 – in the nucleus   **Stage 2** – in the cytoplasm

Unzipped section of DNA

Nuclear membrane

   (a) Name molecules X and Y.

   X _____

   Y _____ **1**

(b) Name organelle Z where the protein chain is being assembled.

1

_____

5. Proteins have many functions. For example proteins can function as enzymes.
   Name one other function of a protein.

1

_____

6. The diagrams below show four stages, A, B, C and D, that occur when an enzyme converts its substrate into a product.

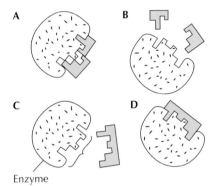

Enzyme

Copy and complete the flow chart by adding letters from the diagram to show the correct order of the stages involved in a **degradation** reaction.

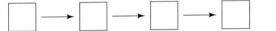

1

7. The diagram below represents the molecular structure of the enzyme catalase.

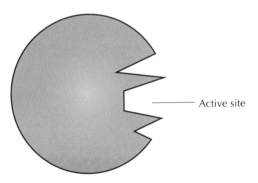

Active site

Describe what happens to the active site when an enzyme is denatured.

_____

1

85

8. Three groups of students investigated the catalase concentration of different tissues.

   Each group set up a test tube containing 5 cm³ of hydrogen peroxide and a cube of tissue. The oxygen was collected over a 5-minute period and the volume was measured as shown in the diagram below.

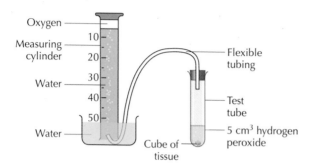

   The procedure was repeated by each group using cubes of liver, apple, carrot and plasticine.

   The results from the three groups are given in the table below.

Tissue	Volume of oxygen collected in 5 minutes (cm³)			
	Group 1	Group 2	Group 3	Average
Liver	44.5	43.5	44	44.0
Apple	1.0	1.5	0.5	1.0
Carrot	4.5	3.0	3.0	3.5
Plasticine	0	0	0	0

   (a) The volume of hydrogen peroxide and time taken to collect the oxygen were kept constant in this investigation.

   State two other variables that must be kept constant.

   1 _____ 1

   2 _____ 1

   (b) State why plasticine was included as a control in this experiment?

   _____

   _____ 1

(c) State what was done in this investigation to make the results reliable.

_____

_____   **1**

(d) Give a valid conclusion that can be drawn from these results.

_____

_____   **1**

9. The graph shows the effect of temperature on the enzyme amylase.

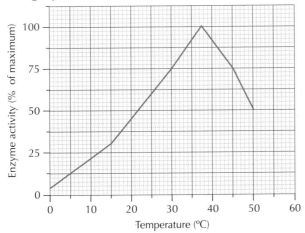

(a) Give the increase in percentage of maximum activity of the enzyme when the temperature was raised from 20 °C to 30 °C.

_____%   **1**

(b) From the graph, predict the temperature at which the enzyme activity will reach zero.

_____°C   **1**

(c) Other than temperature, name one other factor that affects the rate of an enzyme reaction.

_____   **1**

10. An experiment was carried out to investigate the effect of temperature on the rate of respiration of germinating peas. The apparatus was set up as shown below.

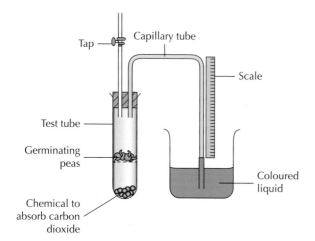

The test tube was placed in a water bath at 5°C. The coloured liquid moved up the capillary tube, as the peas used oxygen for respiration. The distance moved by the coloured liquid was measured against the scale and the rate of respiration per hour was calculated.

The experiment was repeated at different temperatures. The results are shown in the table.

Temperature (°C)	Rate of respiration (cm³ oxygen used per hour)
5	0.5
10	0.25
15	0.40
20	0.70
25	0.95
30	1.20
35	1.45
40	1.25
50	0.20

(a) On a separate piece of graph paper, copy and complete the **line graph** of the results using an appropriate scale to fill most of the graph paper.

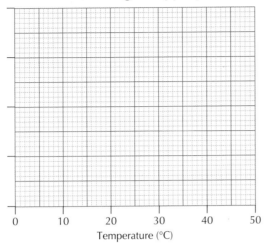

Temperature (°C)

2

11. The diagram below shows aerobic respiration in an animal cell.

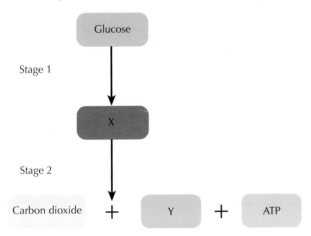

(a) Name substances X and Y.

X _____

Y _____

2

(b) Name the part of the cell where stage 1 takes place.

_____

1

(c) Describe how the reaction would differ in the absence of oxygen.

_____

1

12. During cellular respiration the energy released from the breakdown of glucose is used to generate ATP. The energy transferred by ATP can be used for cellular activities, such as muscle contraction.

Give one other example of a cellular activity that requires ATP.

_____

1

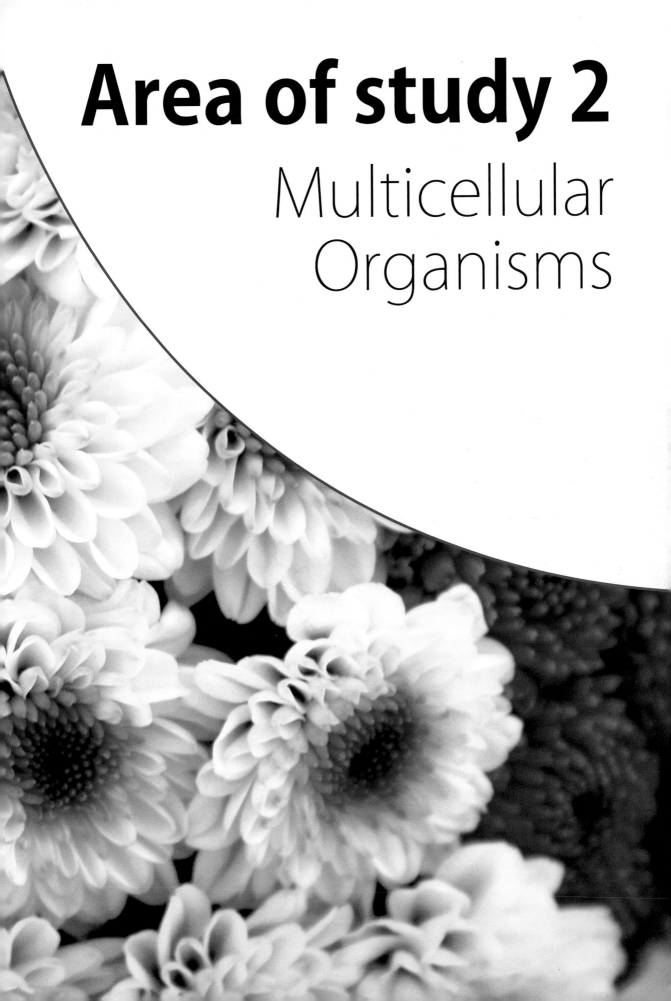

Area of study 2

Multicellular Organisms

7 Producing new cells

You should already know:

- Cells must divide to increase the number of cells in an organism.
- Cell division is controlled by the nucleus.
- During cell division, the parent cell divides to produce two identical cells, which contain the same number of chromosomes in their nuclei as the parent cell.
- Cell division allows organisms to grow and repair damaged parts, e.g. cuts, broken bones.
- Stem cell technology and using cells to grow artificial organs are other examples of therapeutic uses of cells.

Learning intentions

- Describe the sequence of events of mitosis.
- Understand the meaning of the terms 'chromatids', 'equator' and 'spindle fibres'.
- Explain why mitosis is used by cells.
- Describe the maintenance of diploid chromosome complement by mitosis.
- Define the term 'stem cell'.
- Explain where stem cells can be found.
- Describe the role of stem cells in animals.
- Explain the importance of stem cells to animals.
- Explain what is meant by the term 'specialisation' of cells in animals and plants.
- Describe the levels of organisation found in animals and plants.
- Give details of how the structure of an animal or plant cell can relate to its function.

Chromosomes and cells

Within the nucleus of each human body cell are 46 single chromosomes or 23 pairs of chromosomes. One chromosome from each pair is inherited from each parent. The chromosomes contain all of the genetic information.

A karyotype can be produced by analysing the chromosomes found in the nucleus of a cell. This allows for the size, number and shape of the chromosomes to be determined. The human karyotype shown in the diagram shows how the 46 chromosomes are arranged into 23 pairs (fig 7.1).

In figure 7.3 you can see the difference between the male and female karyotypes. A cell that contains a double set of chromosomes is said to be **diploid**.

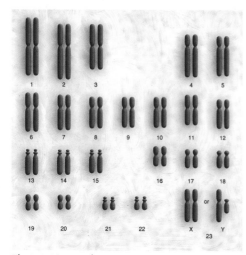

Fig 7.1 *Human karyotype*

In humans a diploid cell contains 46 chromosomes – two matching sets of chromosomes, one from each parent. Nearly all body cells are diploid.

The number of chromosomes found in a cell is also known as the **chromosome complement**.

Sexually-reproducing organisms originate from a single fertilised egg cell. Through cell division, the number of cells is increased. Cells can then become specialised to carry out specific functions in a process called differentiation.

A selection of differentiated cells are shown below (figs 7.4 to 7.7).

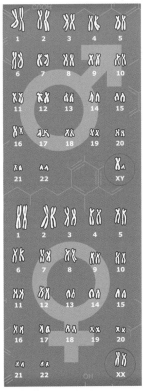

Fig 7.3 *The difference between the male and female karyotypes*

> ### 🌳 Biology in context
>
> The chromosomes shown in karotypes as pair 23 are the sex chromosomes. These determine your sex. Females inherit two X chromosomes, whereas males inherit an X and a Y chromosome (fig 7.2).
>
>
>
> **Fig 7.2** *The human X and Y chromosomes*

Fig 7.4 *Phloem cell*

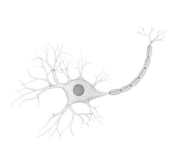

Fig 7.5 *Xylem cell*

Fig 7.6 *Nerve cell*

Fig 7.7 *Red blood cell*

🔬 Make the link – Biology

You will learn more about inheritance in Area of study 2, page **136**.

🔬 Make the link – Biology

You will learn more about different cell types and their functions later on in this chapter, pages **100** and **102**.

GO! Activities

Activity 2.7.1 Working individually

1. Write down full answers to the following questions:
   (a) Describe what a karyotype shows.
   (b) Give the term used to describe a cell with two matching sets of chromosomes.
   (c) Explain what is meant by the term 'chromosome complement'.

2. The information in the table below shows the chromosome complement for different species.

   Using the information in the table, construct a bar chart to show the chromosome complement for each of the different species.

Species	Chromosome complement	Species	Chromosome complement
Lettuce	18	Tomato	24
Cat	38	Human	46
Potato	48	Horse	64
Dog	78		

Activity 2.7.2 Working in groups

Work with a partner to create a model of a human karyotype.

* Cut out 46 chromosomes – 23 in one colour and a matching set in a second colour.
* Make sure that each of the chromosomes in a pair is the same size and shape.
* Match up your 23 pairs and stick them down onto another sheet of paper.
* Label each pair 1 to 23.

You now have your own chromosome complement showing diploid pairs.

Biology in context

The majority of the human population have a chromosome complement of 46 chromosomes. However, some children are born with the genetic condition Down's syndrome.

The condition is named after the British doctor, John Langdon Down, who classified it in 1866. Down's syndrome is a result of the inheritance of an extra copy of one chromosome. This gives a chromosome complement of 47 chromosomes instead of the usual 46.

Fig 7.8 *A child who has Down's syndrome*

In chromosome pair 21, there are three chromosomes instead of two. This extra chromosome causes physical and mental disabilities. In the UK, around 1 out of 1000 children are born with this condition.

Mitosis

The process of cell division is called **mitosis**. It takes place in animal and plant cells in order to increase the number of cells. Mitosis provides new cells for both the growth of cells and the repair of damaged cells.

It is essential that a parent cell replicates the chromosomes contained within its nucleus before it divides. This ensures that the two daughter cells produced can receive the same number of chromosomes as the original cell.

Mitosis can be broken down to a number of stages as shown in figure 7.9.

Hint

Although there are six stages of mitosis shown here, the number of stages shown in diagrams will vary. Make sure you know what you are looking at and can describe what is happening.

 Stage 1 Mitosis starts off with a diploid parent cell. The individual chromosomes coil up and become visible.

 Stage 2 Each chromosome is replicated so that an exact copy of each chromosome is made. The chromosomes consist of two chromatids joined together by a centromere.

 Stage 3 The chromosomes line up at the equator (centre) of the cell, and spindle fibres attach to each pair of chromatids.

 Stage 4 The spindle fibres shorten, pulling the chromatids apart to form new chromosomes which move towards the opposite poles of the cell.

 Stage 5 The nuclear membranes re-form around each group of new chromosomes and the cytoplasm divides.

 Stage 6 This produces two new daughter cells, each with the same number of chromosomes as the original cell.

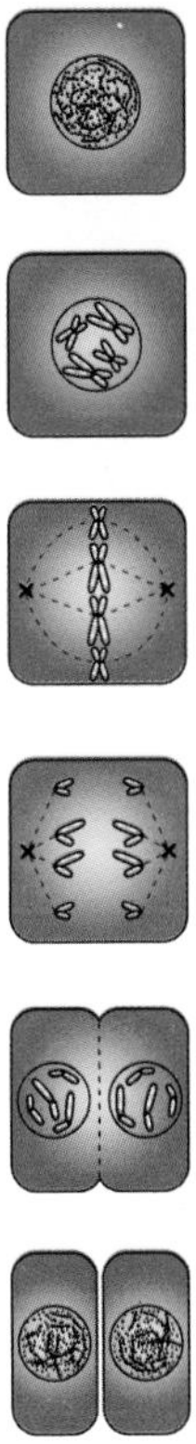

Fig 7.9 *The stages of mitosis*

Maintaining the chromosome complement

Mitosis maintains the chromosome complement so that daughter cells contain the same complement as each other and their parent cell. This ensures that the daughter cells have all the genetic information they need.

What happens to the number of chromosomes is summarised in the simple diagram of cell division shown in figure 7.10.

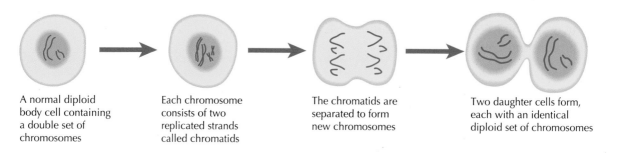

A normal diploid body cell containing a double set of chromosomes

Each chromosome consists of two replicated strands called chromatids

The chromatids are separated to form new chromosomes

Two daughter cells form, each with an identical diploid set of chromosomes

Fig 7.10 *Maintaining the chromosome complement*

🔍 Hint

Remember you can always use the internet to search for short film clips of mitosis to help you understand the process.

📍 Activities

Activity 2.7.3 Working individually

Write down full answers to the following questions:

1. Define mitosis.
2. Describe the stages of mitosis.
3. Describe how a diploid chromosome complement is maintained by mitosis.

Activity 2.7.4 Working in pairs

1. Work with a partner to produce a poster showing each of the stages of mitosis. Base your poster on a cell that has a chromosome complement of 4 chromosomes.

   - Use different colours of pen, drinking straws or pipe cleaners to represent the chromosomes.
   - Use the same colours/size of chromosomes/chromatids in each of the stages to show how mitosis progresses.
   - Add short descriptions to each stage to describe what is happening at each of the stages.

2. Using six small squares of card, draw each stage of mitosis on a separate card. Mix the cards up and practise putting them into the correct order. Make sure that you can describe what happens at each of the stages. Now give them to your partner so that they can try.

I can:

- State that diploid cells are cells that have two matching sets of chromosomes.

- State that human body cells are diploid (contain 46 chromosomes).

- State that mitosis provides new cells for the growth and repair of damaged cells.

- Describe each of the stages of mitosis (including the movement of chromatids to the equator of the cell and the formation of spindle fibres).

- Explain that when a chromosome is duplicated during mitosis it is then described as being made up of two chromatids, with each chromatid being identical.

- Explain that the equator is the centre of the cell, where the chromosomes line up.

- Explain that spindle fibres attach to each pair of chromatids, shortening to pull the chromatids apart.

- State that the process of mitosis maintains the diploid chromosome complement.

- Explain that the diploid chromosome complement is maintained when the chromosomes in diploid cells are replicated during mitosis.

Stem cells

Stem cells (fig 7.11) are found in animals. They are unspecialised cells, which can divide in order to self-renew or can develop into specialised cells. Stem cells have the potential to develop into different types of body cell and are involved in **growth** and **repair** of body tissues. Scientists are currently conducting medical research and developing treatments for certain illnesses that involve the use of stem cells. There are two main types of stem cells: embryonic stem cells and adult stem cells.

Embryonic stem cells are found in embryos. They can be obtained from the embryo at a very early stage and have the ability to develop into any type of body cell. They allow an organism to grow from a tiny embryo to a fully formed individual. Research on the possible use of embryonic stem cells in medicine is underway but there are ethical issues involved.

Tissue stem cells (or adult stem cells) can be found in the body throughout life. Despite being called adult stem cells, they can be found in children as well as adults. Tissue stem cells do not have the same properties as embryonic stem cells. For example tissue stem cells found in the bone marrow are only capable of becoming a variety of different blood cell types. It is thought that tissue stem cells can only produce the cell types of the tissue in which they are found.

Stem cells have important **medical applications**. The table below shows the current uses of stem cells in medicine.

Fig 7.11 *A coloured high-powered microscope image of stem cells*

Fig 7.12 *Skin grown in a lab from stem cells is sliced and stretched before being used to treat burn victims*

Source of stem cell	Medical use
Marrow in the centre of bones	Treating leukaemia, a type of cancer caused by abnormal blood cells.
Skin	Growing new layers of skin that can be used to treat burn victims (fig 7.12).
Heart muscle	Repairing damaged heart muscle after a heart attack (this technique is still being tested).
Bladder	Building a new bladder in a laboratory for a patient whose bladder has been damaged by injury or disease (fig 7.13).

Fig 7.13 *Cross section through a bladder. The bladder is a hollow organ and is therefore easier to build than an organ such as the heart*

In the future, scientists hope that stem cells will be used to cure conditions such as Alzheimer's disease, diabetes, spinal-cord injuries and stroke. Stem cells may also provide a useful alternative to animals for testing experimental drugs.

♈ Biology in context

There are many issues surrounding the use of stem cells, in particular embryonic stem cells.

- Research using embryonic stem cells involves the destruction of embryos. Many people believe that it is unethical to destroy embryos because they have the potential to grow into a baby.
- Some people believe we should use only adult stem cells because these do not require the destruction of embryos.
- Some of the techniques used in stem-cell research are closely related to cloning, which many people oppose.
- Like any new medical advance, there are also safety concerns surrounding the use of stem cells. For example, some research has suggested that transplanted stem cells could become cancerous.

GO! Activity

Activity 2.7.5 Working individually

Write down full answers to the following questions:

1. State where stem cells are found.
2. Give two properties of stem cells.
3. Name the two main types of stem cell.
4. Describe the difference between the properties of the two types of stem cells.

Activity 2.7.6 Working individually

Carry out some research into stem cells obtained from embryos. Write an essay that discusses the use of embryonic stem cells. Your essay should be laid out in the following format:

Introduction – introduce the issue
Arguments for – give two or three arguments for using stem cells obtained from embryos
Arguments against – give two or three arguments against using stem cells obtained from embryos
Conclusion – summarise your essay and give your own opinion.

I can:

- State that stem cells in animals are unspecialised cells.
- State that stem cells can divide in order to self-renew.
- State that stem cells have the potential to become different types of cells.
- Describe the role of stem cells in animals as the cells that give rise to specialised cells.

(Continued)

- State that stem cells are involved in growth and repair.

- Explain that stem cells can be obtained from the embryo at a very early stage.

- Explain that tissue stem cells can be found in the body throughout life.

Specialisation of cells

Multicellular organisms have more than one cell type and are made up of tissues and organs. Organs (e.g. heart, lungs) perform different functions. The cells in organs are specialised for their function and work together to form systems.

Similar specialised cells working together form a tissue. Groups of tissues form organs. Organs work together in groups called systems.

The simple flow diagram below demonstrates the hierarchy that exists, showing how basic cells eventually lead to the formation of systems.

$$\text{Cells} \rightarrow \text{tissues} \rightarrow \text{organs} \rightarrow \text{systems}$$

Specialisation in animal cells

Animal cells show **specialisation**. This means they have a **special shape or structure** that allows them to carry out a **specific function**. The table below shows some animal cells and explains how their structure is related to their function.

Cell	How cell structure relates to function
	Red blood cells carry oxygen around the body. They contain haemoglobin to do this. Red blood cells do not contain a nucleus or ribosomes, so they can contain as much haemoglobin as possible. Red blood cells have a special biconcave shape, which increases their surface area so they can absorb oxygen quickly and efficiently.
	Muscle cells contract and relax to allow movement. They are long and thin, and contain special contractile proteins called myofilaments, which allow them to become shorter, creating muscle contraction.

Cell	How cell structure relates to function
	Neurons send messages around the body as electrical impulses. They are long and thin, and have many projections that allow them to communicate with many other neurons.

Levels of organisation

Animals have levels of organisation in their bodies that allow them to function efficiently. Specialised cells that perform specific functions group together to form **tissues**. For example, muscle cells group together to form muscle tissue. Tissues work together to form **organs** with specific functions. An example is the heart, which contains muscle and nervous tissue. The cells in organs are specialised for their function. For example, the muscle cells in the heart are able to contract in a regular pattern. This allows the heart to beat and pump blood around the body. Organs work together forming a **system**. The heart is part of the circulatory system, which also includes the blood vessels (fig 7.14).

Make the link – Biology

Body cells have many different functions (see pages **100, 110** and **125**).

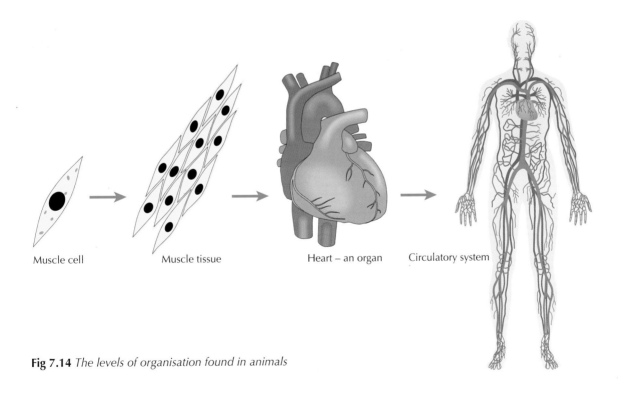

Muscle cell   Muscle tissue   Heart – an organ   Circulatory system

Fig 7.14 *The levels of organisation found in animals*

Activities

Activity 2.7.7 Working individually

1. Explain the term 'specialisation'.
2. Describe how a red blood cell is specialised for its function.
3. Put the following structures in order from the lowest to the highest in the hierarchy:

   cell   system   organ   tissue

Activity 2.7.8 Working individually

Make an 'Animal cell owner's manual'. Cut two sheets of A4 paper in half and arrange them into a booklet, stapling the centre together. On each page of the booklet draw a diagram of a different type of animal cell and explain how its structure is related to its function.

Make the link – Biology

Look through Chapter 11. Can you identify some of the plant cells shown here? Can you relate their structure to their function?

Specialisation in plant cells

Like animal cells, plant cells show **specialisation**. Plant cells are adapted to carry out many different functions. The table below shows some plant cells and explains how their structure is related to their function.

Cell	How cell structure relates to function
	**Root hair cells** absorb water. They have a large projection that increases their surface area for absorbing water.
	Palisade mesophyll cells are found in leaves. They contain a large number of chloroplasts to allow them to carry out photosynthesis efficiently.

Cell	How cell structure relates to function
	Xylem cells transport water in plants. Xylem cells take the form of hollow tubes that allow water to travel from the roots to the leaves in a continuous stream.
	Phloem cells transport food in plants. The end walls of phloem cells are like sieves, which allow glucose to pass from one cell to the next.
Closed stoma Open stoma Pore closed Pore open	Plant leaves have pores called stomata, which allow gases to enter and leave the leaf. **Guard cells** change shape to open or close the pore.

Levels of organisation in plants

Like animals, plants have different levels of **organisation**. For example xylem **cells**, which transport water, join together to form xylem **tissue**. Xylem tissue works alongside phloem tissue in an **organ** called the vascular bundle. The cells in plant organs are specialised to carry out a specific function. For example the phloem cells in the vascular bundle have end walls like sieves that allow sugar to pass easily from one cell to the next. The vascular bundle is part of the vascular **system** found in plants.

Activities

Activity 2.7.9 Working as a group

For this activity you will need to split your group into teams. Each team must then choose an organism to research: either humans or plants.

Each team needs to copy out the hierarchy of organisation flow diagram at the top of a piece of paper.

E.g. Cells → tissues → organs → systems

Using any resources available (e.g. books, Internet, etc.) each team must list as many examples as possible of the different types of cells, tissues, organs and systems found in their chosen organism. The team with the greatest number of correct examples wins.

Here is an example of each to get you started.

Humans
Muscle cells → muscle tissue → heart → circulatory system

Plants
Root hair cell → root tissue → root → vascular bundles in transport system

Activity 2.7.10 Working individually

Make a 'Plant cell owner's manual'. Cut two sheets of A4 paper in half and arrange them into a booklet, stapling the centre. On each page of the booklet draw a diagram of a different type of plant cell and explain how its structure is related to its function.

I can:

- Explain that specialisation of cells leads to the formation of a variety of cells, tissues and organs. ⬭ ⬭ ⬭

- State that multicellular organisms have more than one cell type and are made up of tissues and organs. ⬭ ⬭ ⬭

- State that groups of organs work together to form systems. ⬭ ⬭ ⬭

- State that organs (e.g. heart, lungs) perform different functions. ⬭ ⬭ ⬭

- State that the cells in organs are specialised for their function and work together to form systems. ⬭ ⬭ ⬭

- Describe the levels of organisation seen in multicellular organisms, from groups of cells forming a tissue, tissues forming organs and organs working together to form a system. ⬭ ⬭ ⬭

- Draw a simple flow diagram to show the hierarchy of organisation in multicellular organisms. ⬭ ⬭ ⬭

    E.g. Cells → tissues → organs → systems

8 Control and communication

You should already know:

- The structure and function of key organs and organ systems.
- For example the nervous system, which allows messages to be sent to and from the brain.

Learning intentions

- Describe the structure of the brain.
- Give details of the functions of the different structures of the brain.
- Describe the structure of the nervous system.
- Describe the structure and function of the central nervous system (CNS).
- Name the three types of neurons.
- Describe the role of sensory, inter and motor neurons.
- Describe the role of receptors.
- Explain how messages are carried along neurons.
- Describe the different responses brought about by the CNS.
- Explain what a reflex action is and give some examples.
- Explain the importance of reflex actions.
- Explain the role of sensory, inter and motor neurons in a reflex arc.
- Explain the importance of synapses in the nervous system.
- State the function of endocrine glands.
- Describe the role of hormones in signalling.
- Explain how hormones target specific tissues.
- Describe how blood glucose is regulated.
- Describe the roles of insulin, glucagon, glycogen, the pancreas and the liver in controlling blood glucose levels.
- Describe the causes and treatment of type 1 and type 2 diabetes.

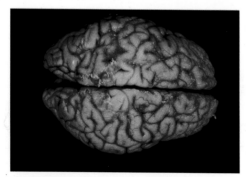

Fig 8.1 *The outer surface of the brain seen from above*

The brain

The brain is the most complex organ in your body. It controls vital processes that keep you alive, enables you to interact with the world around you and makes memories that you can keep for decades. Your brain is capable of doing several things at once. For example right now you are reading, sitting upright, breathing, and your heart is beating. All of these processes require input from your brain.

The brain is organised into distinct areas, each of which carries out a different function. The largest part of the brain is the **cerebrum**. The cerebrum is the area of the brain that enables conscious thought and memory. The cerebrum also receives information from the senses, processes the information and brings about a response. The **cerebellum** is found at the back of the brain, tucked underneath the cerebrum. The cerebellum is responsible for controlling balance and co-ordination. The **medulla** lies above the spinal cord. The medulla is the region of the brain that controls breathing and heart rates.

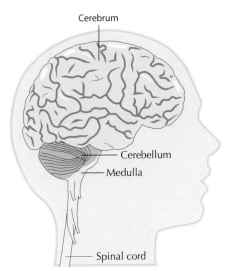

Fig 8.2 *The brain*

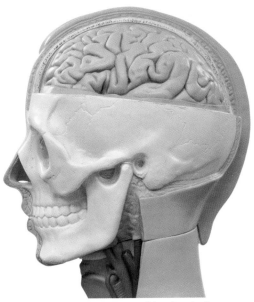

Fig 8.4 *The brain is protected by the skull*

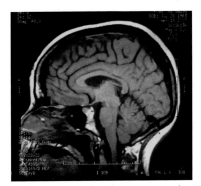

Fig 8.3 *A scan called an MRI can be used to produce an image of the brain*

🌳 Biology in context

Sleep is extremely important for the functioning of the brain. People who experience a period of time without sleep have difficulty concentrating and a reduced attention span. If a period of sleeplessness continues, memory, planning and language can also be affected. On average, adult humans require 8 hours of sleep per night. Other animals require varying amounts of sleep – from a giraffe, which needs just 2 hours, to a tiger, which needs around 16 hours.

Scientists are unsure exactly why we need to sleep but several theories have been put forward:

- Sleep allows the body to repair itself from damage caused through the day.
- Sleep allows the body to replenish its store of ATP.
- Sleep allows your brain to replay events of the day to cement them in your memories.

💥 Make the link – Expressive Arts

In art you practise still-life drawing. Early anatomists often used artists to record their work, before cameras were widely available.

💥 Make the link – Biology

The brain is part of the nervous system (see page **109**).

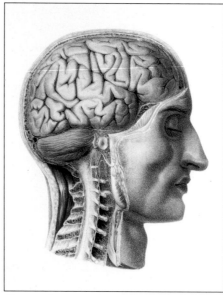

Fig 8.5 *This artwork was completed for an anatomy book published in the mid-1800s*

GO! **Activities**

Activity 2.8.1 Working individually

1. Copy and complete the table below into your notes.

Structure	Function
Cerebrum	
	Controls heart rate and breathing rate
Cerebellum	

Activity 2.8.2 Working in pairs

The 'Stroop effect' was named after a scientist called J. Ridley Stroop, who carried out this experiment in the 1930s. Look at the Stroop effect picture and try to name the colours of the different words. Do not read the word itself, try to name its colour. Try one set at a time.

Which set did you find easier?

Why might this be?

Would this test be easier for young children who know their colours but cannot read yet?

Red   Green
Blue   Red
Green   Blue
Black   Black
Green   Blue
Blue   Red

Fig 8.6 *The Stroop effect*

I can:

- Describe and identify the structures of the brain as the cerebrum, cerebellum and medulla.
- State that the cerebrum enables conscious thought and memory.
- State that the cerebellum controls co-ordination and balance.
- State that the medulla controls heart and breathing rates.

The nervous system

The nervous system is made up of the brain, spinal cord and other nerves. This is a communication system that uses electrical impulses to send messages from one part of the body to another. This allows communication between the cells of a multicellular organism. Neurons (nerve cells) send messages as electrical impulses from one part of the nervous system to another. Neurons are specially structured to allow them to send signals quickly and efficiently. The central nervous system (CNS) is made up of the **brain** and the **spinal cord** only.

The sense organs contain **receptor cells** which detect sensory input from our surroundings. This information is the stimuli (changes in conditions) that triggers an electrical impulse. Electrical impulses carry messages along neurons. The impulse passes along the **sensory neuron**, which passes this information to the CNS. **Inter neurons** operate within the CNS, which processes the information from the senses that require a response. An electrical impulse is then sent along a **motor neuron** to enable a response to occur at an **effector** (muscle or gland). A response to a stimulus can be a rapid action from a muscle or a slower response from a gland.

One example of nervous-system communication is shown in figure 8.8. The nervous system brings about changes when there is a drop in the external temperature of the body. Shivering is a faster response because it is brought about by muscles. The reduction in sweating occurs more slowly because this is brought about by a gland.

Make the link – Biology

A neuron is an example of a cell that is well suited to carrying out its function (see page **100**).

Make the link – Sciences

In physics you study electrical charges. Electrical impulses allow messages to be sent around the body. Some organisms such as the electric eel use electrical charges for other reasons such as defence.

Hint

Remember the CNS is the brain and spinal cord only. Think about a diagram of the body facing you. The brain and spinal cord are in the centre.

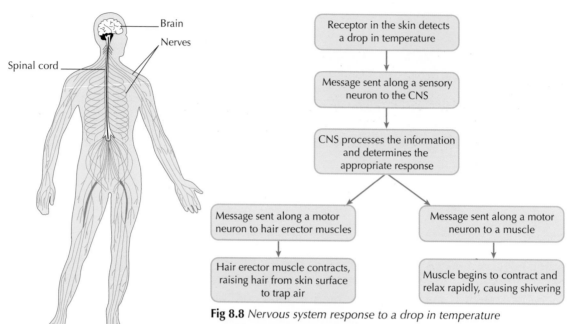

Fig 8.7 *The nervous system*

Receptor in the skin detects a drop in temperature

↓

Message sent along a sensory neuron to the CNS

↓

CNS processes the information and determines the appropriate response

↓ (branches into two)

Message sent along a motor neuron to hair erector muscles

↓

Hair erector muscle contracts, raising hair from skin surface to trap air

Message sent along a motor neuron to a muscle

↓

Muscle begins to contract and relax rapidly, causing shivering

Fig 8.8 *Nervous system response to a drop in temperature*

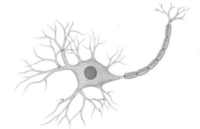

Fig 8.9 *A neuron*

Fig 8.10 *Neurons found in the brain*

◉ Investigation

Reaction time measures the length of time it takes your nervous system to respond to a stimulus. This can be measured using a ruler drop test. The subject holds their thumb and fingers either side of a ruler. The person carrying out the test drops the ruler, and the subject tries to catch it as quickly as possible. Their reaction time is measured as the point on the ruler scale where their hand catches it. The shorter the distance, the faster the reaction time.

Planning and designing an investigation

Design an experiment that would allow you to investigate the effect of age on reaction time.

What factors would need to be kept the same in this investigation?

Making a prediction

How do you think age will affect reaction time? Write down your hypothesis.

Fig 8.11 *Ruler drop reaction-time test*

Results

Here are some sample results from an experiment to investigate the effect of distraction on reaction rate.

Distraction level	Attempt 1 (distance on ruler scale, mm)	Attempt 2 (distance on ruler scale, mm)	Attempt 3 (distance on ruler scale, mm)	Average (distance on ruler scale, mm)
No distraction	195	175	155	
Quiet music	205	170	165	
Moderate music	200	180	190	
Loud music	230	215	215	

Calculate the average reaction time at each level of distraction.

Draw a bar graph to display the average results.

Conclusion

What do these results show?

What is the relationship between distraction and reaction rate?

Can you explain these results?

Evaluation

How could the reliability of this experiment be improved?

GO! Activities

Activity 2.8.3 Working individually

1. Draw a diagram of the nervous system and label the brain, spinal cord and nerves.
2. Name the two structures which make up the central nervous system.
3. Describe the role of sensory and motor neurons.
4. Copy and complete the sentence below, selecting the correct option from each bracket.

   Effectors bring about responses. Muscles bring about (rapid/slower) responses whereas glands bring about (rapid/slower) responses.

Activity 2.8.4 Working in a group

This should be carried out under the supervision of a teacher. Set up three bowls of water. The left bowl should contain cold water with ice cubes, the middle should be left at room temperature, and the right bowl should contain hot (not burning hot!) water. Place your left hand in the left bowl and right hand in the right bowl for a few minutes. Then move both hands into the middle bowl at the same time.

   What do you feel?

   Is there a difference between your left and right hands?

   Why might this be?

(Continued)

Activity 2.8.5 Working in pairs

Carry out the ruler drop test as described above, using sight to test reaction time. Try it a second time with the subject blindfolded. The person carrying out the test should say 'Go' when they drop the ruler this time.

Was there a difference in reaction time when using sound rather than sight as a stimulus? Why might this be?

I can:

- State that the nervous system consists of the central nervous system (CNS) and other nerves.
- State that the CNS is made up of the brain and the spinal cord.
- State that a stimulus is a change in conditions and is detected by receptors.
- State that there are three types of neurons: sensory, inter and motor.
- State that electrical impulses carry messages along neurons.
- Describe the role of sensory neurons as passing information (in the form of an electrical impulse) to the CNS.
- Describe the role of inter neurons as passing information (in the form of an electrical impulse) from the CNS to the motor neurons.
- Describe the role of motor neurons as enabling a response to occur at an effector (muscle or gland).
- State that a response to a stimulus can be a rapid action from a muscle or a slower response from a gland.

Reflex action

Reflex actions are fast, **automatic** responses. They occur in response to a specific stimulus and usually have a **protective** role. Reflex actions happen without any input from the brain and therefore cannot be controlled.

For example, if your hand touches a very hot object, it will automatically pull away. This happens very **rapidly** to remove the hand from danger as quickly as possible. The table on the next page shows some examples of reflex actions. Our reflexes protect the body from harm. For example the gag reflex helps prevent choking. The pupillary light reflex protects the sensitive cells at the back of the eye from damage.

 Hint

Remember to RAP a Reflex:

R – rapid
A – automatic
P – protective

Name of reflex action	Stimulus	Response	Advantage
Pupillary light reflex	Bright light	Pupils become smaller	Cells at the back of the eye are protected from high light intensity
Blink reflex	Touching the eye	Eyelids blink	Protects the eye from potential damage
Gag reflex	Touching the back of the tongue or throat	Muscles in the back of the throat contract, causing gagging	Helps to prevent choking
Pharyngeal swallow reflex	Food touching the back of the throat	Causes swallowing	Helps to prevent choking
Mammalian diving reflex	Cold water touching the face	Heart rate slows	Allows survival underwater for longer periods of time

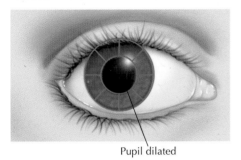 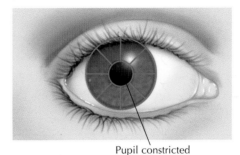

Pupil dilated

Pupil constricted

Fig 8.12 *Left: the response of the eye to dim light. Right: the response of the eye to bright light*

🌳 **Biology in context**

Some reflex actions are only shown by infants. These actions gradually decrease and disappear after a few months. The 'Moro reflex' occurs when a baby feels as if it is falling. This reflex action causes the arms to spread out and then move back in towards the body.

All newborn mammals show a suckling reflex. They will suck anything that touches the roof of their mouth. These reflexes help to protect infants from harm. For example the Moro reflex may have evolved to help infants maintain a grasp on their mother.

Fig 8.13 *Babies show a palmar grasp reflex, which causes them to grasp anything that touches their palm*

GO! Activities

Activity 2.8.6 Working individually

1. Give the meaning of the term 'reflex action'.
2. Describe the importance of reflex actions.
3. Describe the pupillary light reflex and its importance.
4. Describe the mammalian diving reflex and explain how it benefits these animals.

Activity 2.8.7 Working in pairs

1. The patellar reflex action involves striking the leg just below the knee cap. This causes the lower leg to kick upwards.

   - Work with a partner.
   - Ask your partner to sit with one leg on top of the other and relax.
   - Using the edge of your hand, strike your partner's knee gently, just below the knee cap and watch for a response.

Fig 8.14 *Doctors bring about the knee-jerk reaction using a reflex hammer*

2. Ask your partner to close their eyes for 30 seconds. After 30 seconds tell them to open their eyes. Look for the pupillary reflex response.

Reflex arc

Reflex actions do not require conscious thought from the brain. However, in most cases, impulses sent to the brain give awareness that the reflex is happening. During a reflex action, **electrical impulses** are sent from a receptor to an effector through a series of neurons. This pathway of neurons is called a **reflex arc**.

If you accidentally touch a hot object, this stimulus is detected by **receptors** in the skin. This causes an electrical impulse to be sent along a **sensory neuron**. The electrical impulse is passed on to an **inter neuron** and then to a **motor neuron**. The motor neuron is connected to an **effector**, in this case a muscle. The impulse causes the muscle to **contract** and moves your hand away from damaging heat, protecting the body from harm. This process is summarised below:

stimulus → receptor → sensory neuron → inter neuron → motor neuron → effector → response

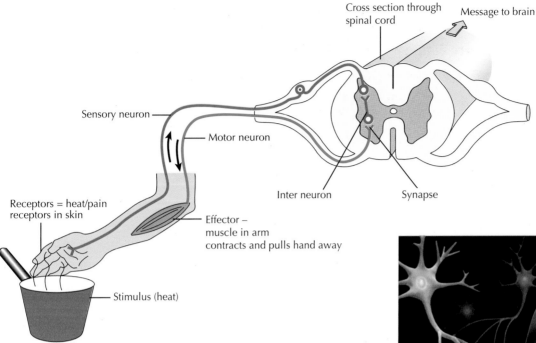

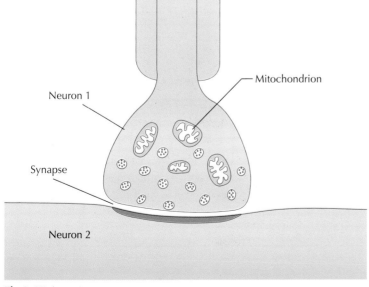

Fig 8.15 *A reflex arc*

In a reflex arc, electrical impulses move from one neuron to another. However, neurons are not in direct contact with each other. Between two neurons there is a small gap called a **synapse**. The electrical impulse from one neuron is passed to the next using **chemicals**. These chemicals (known as neurotransmitters) are released from one neuron and pass across the synapse. They bind onto receptors and initiate an electrical impulse in the second neuron. This process allows electrical impulses to pass from one neuron to another throughout the nervous system.

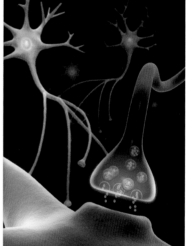

Fig 8.16 *Chemicals called neurotransmitters move from one neuron to another across a synapse*

Fig 8.17 *A synapse*

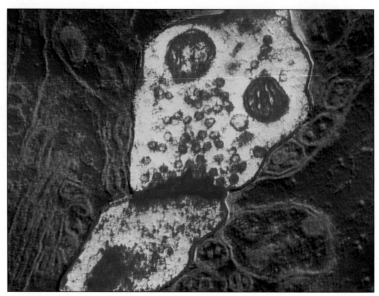

Fig 8.18 *Colour has been added to this electron microscope image. The synapse appears as a dark red line in the centre of the photo*

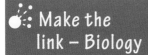

Make the link – Biology

A stimulus is a change in conditions and is detected by receptors (see page **109**).

GO! Activities

Activity 2.8.8 Working individually

1. Describe the flow of information along a reflex arc.
2. Name the small gaps found between neurons.
3. Explain the role of these small gaps.

Activity 2.8.9 Working in pairs

Make a slideshow presentation to inform someone of your own age about reflex responses and the reflex arc.
Your presentation should include:

- What a reflex action is and some examples.
- An explanation of the terms 'stimulus', 'receptor' and 'effector'.
- Information about a reflex arc including the names of the neurons involved.
- What a synapse is and why it is important.

Activity 2.8.10 Working in pairs

Make a 10-question quiz on reflex actions and the reflex arc. Try to include a mixture of easy and challenging questions. Swap your quiz with a partner and test each other's knowledge.

I can:

- State that reflexes protect the body from harm.
- Describe a reflex arc as the passage of an impulse from a sensory neuron, across an inter neuron to a motor neuron.
- State that electrical impulses carry messages along neurons.
- Describe a synapse as a gap between neurons that allows chemicals to transfer messages from one neuron to another.

Hormones

Multicellular organisms use chemical messengers called **hormones** to send **messages** from one part of the body to another. Hormones are produced by **endocrine glands** and released into the bloodstream. In the bloodstream they can travel to other parts of the body where they have their effect. The tissue a hormone has its effect on is called a **target tissue**.

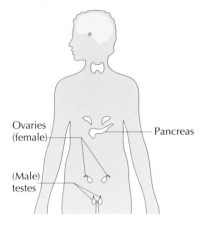

Ovaries (female)

Pancreas

(Male) testes

Fig 8.19 *The position of some endocrine glands within the body*

Make the link – Biology

Other molecules such as oxygen and carbon dioxide are transported in the blood (see page **172**).

Target tissues have cells with special receptor proteins on their surface. When a hormone binds onto the receptor protein it brings about changes in the cell. The receptor proteins are complementary in shape to specific hormones. This ensures that only tissues with receptor proteins for a specific hormone will be affected by it.

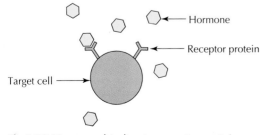

Hormone

Receptor protein

Target cell

Fig 8.20 *Hormone binding to receptor protein*

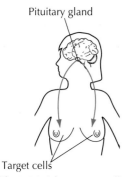

Pituitary gland

Target cells

Fig 8.21 *A hormone called prolactin is produced in the brain and travels to the breasts where it stimulates milk production in women who have just given birth*

Biology in context

The hormones oestrogen and testosterone are often thought of as male and female hormones. However, both hormones are found in both sexes. Males produce more testosterone than women, and women produce more oestrogen than men.

In women, oestrogen is mainly produced by the ovaries and has many target tissues including the uterus and breasts. Oestrogen is involved in breast development and regulating the menstrual cycle. In men, oestrogen is required for the testes to produce fully functional sperm.

Men produce testosterone in the testes. It is involved in the growth of the penis during puberty, body and facial hair growth, and the deepening of the voice, and is necessary for normal sperm development. In women, testosterone is produced by the ovaries. It is involved in maintaining muscle mass and bone strength.

Make the link – Health and Wellbeing

Some body builders use testosterone illegally to help increase muscle bulk. This can result in bigger muscles but can also lead to high blood pressure, heart problems or liver damage if used for long periods of time.

GO! Activities

Activity 2.8.11 Working individually

1. Describe the role of hormones in the body.
2. Name the glands that release hormones.
3. Describe how a hormone travels from where it is produced to where it has an effect.
4. Explain why hormones will act only on certain target tissues.

Activity 2.8.12 Working in pairs

Draw or trace an outline of the human body onto an A4 piece of paper.

Research the following hormones:

- ADH
- Adrenaline
- Human growth hormone

On your outline, use a different colour to mark where each of these hormones is made and where it has its effect. Draw a key to show which hormone is represented by each colour.

I can:

- State that hormones are chemical messengers.
- State that endocrine glands release hormones into the bloodstream where they travel to target tissues.
- State that target tissues have cells with complementary receptor proteins for specific hormones.
- Explain that only tissues with a corresponding receptor will be affected by a specific hormone.

Control of blood glucose levels

Animals gain most of their energy by respiring **glucose**. This means the cells of the body require a constant supply of glucose to release energy. Glucose is transported around the body in the **blood**. The body uses control mechanisms to keep blood glucose levels constant.

After eating a meal, blood glucose levels increase. This rise in blood glucose levels is detected by the **pancreas**, which increases the release of a hormone called **insulin**. Insulin is released into the blood and travels to the **liver**. Insulin binds to **receptor proteins** on the cells of the liver and activates enzymes which convert **glucose** into **glycogen**. Glycogen acts as a carbohydrate store and this process brings blood glucose levels back down to their normal set point.

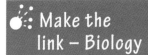
Make the link – Biology

Look back at Chapter 6, as a reminder of how energy is released from the breakdown of glucose. Read about the transport of glucose through the body in Chapter 12 and how it is absorbed into the blood in Chapter 13.

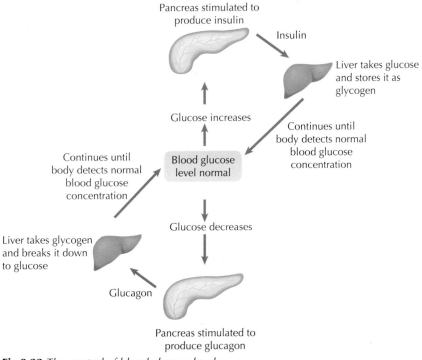

Pancreas stimulated to produce insulin

Insulin

Liver takes glucose and stores it as glycogen

Glucose increases

Continues until body detects normal blood glucose concentration

Continues until body detects normal blood glucose concentration

Blood glucose level normal

Glucose decreases

Liver takes glycogen and breaks it down to glucose

Glucagon

Pancreas stimulated to produce glucagon

Fig 8.22 *The control of blood glucose levels*

In between meals, blood glucose levels decrease. This change in blood glucose levels is detected by the pancreas. In this case, the pancreas releases a hormone called **glucagon** into the blood. Glucagon travels to the **liver** where it binds to receptor proteins on the cells of the liver and activates enzymes that convert **glycogen** into **glucose**. This raises the blood glucose level back up to its normal set point.

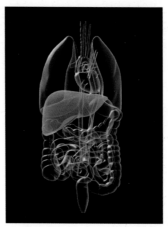 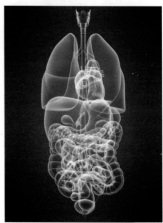

Fig 8.23 *The liver and pancreas are involved in controlling blood glucose levels*

Make the link – Biology

People who suffer from diabetes cannot regulate their blood glucose levels (see page **121**).

All cells require a supply of glucose to carry out respiration (see page **68**).

GO! Activities

Activity 2.8.13 Working individually

1. Name the organ that detects changes in blood glucose levels.
2. Name the carbohydrate that is used to store glucose.
3. Describe the role of insulin in controlling blood glucose levels.
4. Describe the role of glucagon in controlling blood glucose levels.

Activity 2.8.14 Working individually

As well as acting on the liver, insulin has other target tissues. Carry out some research into the role of insulin and write a 100-word report on your findings.

Activity 2.8.15 Working in pairs

With a partner, write down each of the following stages in controlling blood glucose levels onto small pieces of card or sticky notes:

- Normal blood glucose levels
- Blood glucose levels increase
- Increase in blood glucose levels detected by the pancreas
- Pancreas releases insulin into the blood
- Insulin travels to the liver
- Insulin brings about the change of glucose into glycogen
- Blood glucose levels decrease back to set point

- Blood glucose levels decrease
- Decrease in blood glucose levels detected by the pancreas
- Pancreas releases glucagon into the blood
- Glucagon travels to the liver
- Glucagon brings about the change of glycogen into glucose
- Blood glucose levels increase back to set point

When you are finished, put the cards in the correct order to show the response to a decrease in blood glucose levels. Do the same for an increase in blood glucose.

I can:

- State that a change in blood glucose levels is detected by the pancreas.
- State that glucose can be stored as glycogen in the liver.
- Explain that when blood glucose level increases, the pancreas releases more insulin, which travels to the liver where it activates enzymes in the liver cells to convert glucose into glycogen.
- Explain that when the blood glucose level decreases, the pancreas releases more glucagon, which travels to the liver where it activates enzymes in the liver cells to convert glycogen into glucose.

Diabetes

Nearly a quarter of a million people in Scotland suffer from **diabetes**. This number has been increasing year on year and is largely due to type 2 diabetes rather than type 1.

People who suffer from **type 1 diabetes** do not produce enough insulin. In some cases they produce no insulin at all. This means they have a high blood glucose level, especially after a meal. Insulin signals body cells to absorb glucose, which can be used to produce energy. In people who suffer from type 1 diabetes this can no longer happen, and the glucose remains in the bloodstream. Type 1 diabetes is treated by injecting insulin regularly throughout the day. Before injecting insulin, a person with type 1 diabetes must check their blood glucose level to ensure they are injecting the correct quantity of insulin.

In people suffering from **type 2 diabetes**, their body cells no longer respond to insulin. This is sometimes known as insulin

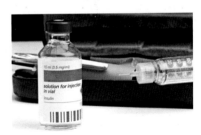

Fig 8.24 *Insulin is made by genetic engineering*

Fig 8.25 *Many diabetics use this pen-like device to deliver insulin rather than a syringe*

resistance because the body is capable of producing insulin, however, it does not have any effect on the cells. Like type 1 diabetes, this causes high blood glucose levels. Type 2 diabetes is treated using lifestyle changes. For example eating healthily, exercising regularly and losing weight if necessary. If lifestyle changes alone do not work, medications that lower blood glucose levels may also be given.

It is important that people with both type 1 and type 2 diabetes control their blood glucose levels properly. Left unchecked, high blood glucose levels can cause damage to delicate blood vessels, especially those in the eyes and kidneys. If the blood glucose level remains high over a long period of time this can damage vision or cause kidney failure.

People who have undiagnosed diabetes often experience a symptom called polyuria. This means they urinate more frequently than normal. This symptom is caused by the high levels of glucose in the blood. This high glucose level affects osmosis and draws water into the urine, causing large volumes of urine to be produced. High blood glucose levels can also cause blurry vision. This is caused by high glucose levels in the lens of the eye, which draws water inside.

Make the link – Biology

Insulin is made by genetic engineering (see page **65**).

Osmosis is the diffusion of water (see page **33**).

Fig 8.26 *People with type 1 diabetes must check their blood sugar regularly*

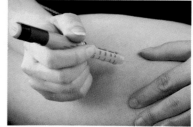
Fig 8.27 *Insulin is injected under the skin (subcutaneously), usually in the leg or stomach*

Biology in context

In some cases, diabetes is diagnosed using a glucose tolerance test (GTT). The patient must not eat or drink for around 12 hours before the test. They are given a measured quantity of glucose and blood samples are taken from them every 30 minutes for 2–3 hours.

A person without diabetes will show a rise in blood glucose level after the glucose is ingested and a quick decrease as the pancreas produces insulin to bring the blood glucose level back down to normal. A person with diabetes will have a high initial blood glucose level which increases further after taking the glucose and remains high for several hours.

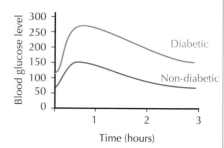
Fig 8.28 *A graph comparing the GTT results of a person with diabetes and a person without diabetes*

GO! Activities

Activity 2.8.16 Working individually

1. Describe the cause of type 1 diabetes.
2. Describe the cause of type 2 diabetes.
3. Explain the differences between the treatment of type 1 and type 2 diabetes.

Activity 2.8.17 Working in pairs

Frederick Grant Banting and John James Rickard Macleod were awarded the Nobel prize in physiology (medicine) in 1923 for the discovery of insulin. Carry out some research to answer the following questions:

- How was diabetes treated before the discovery of insulin?
- What experiments did the scientists carry out to investigate the function of the pancreas?
- Who was Charles Herbert Best?
- The experiments these scientists carried out involved the use of dogs, but their discoveries have saved countless lives. How do you feel about using animals for experiments like this? Try to think about both sides of the argument.

Activity 2.8.18 Working in pairs

Make an information leaflet or a slideshow presentation that could be given to someone who has just been diagnosed with type 1 or type 2 diabetes. Include information about:

- the cause of their disease
- symptoms of diabetes
- how their disease is treated
- the dangers associated with high blood glucose levels.

Make the link – Health and Wellbeing

Diabetes is a national problem in Scotland. Type 2 diabetes is often caused by lack of exercise and/or poor diet.

9 Reproduction

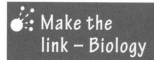
Make the link – Biology

Look back to Chapter 7 as a reminder of the importance of diploid cells.

Fig 9.1 *Sexual reproduction produces offspring that are similar but not identical to their parents*

Reproduction

All living organisms must have a method of reproduction to produce new individuals and continue their species. Most animals use **sexual reproduction** to achieve this. Sexual reproduction involves combining genetic information from two individuals (parents) to produce offspring. The offspring produced by sexual reproduction are similar to their parents but not identical (fig 9.1).

During sexual reproduction, sex cells from each parent fuse together to form a **zygote**. Sex cells are known as **gametes**. Gametes are **haploid**. This means they have only one set of chromosomes. When the male and female gametes fuse together, the zygote produced is **diploid**. This means it has two sets of chromosomes.

Sex cells in humans

In animals the male gametes are sperm cells. A sperm cell has a head section containing the nucleus and a tail section that allows it to swim. A sperm cell also has many mitochondria to provide the energy required for movement (fig 9.2).

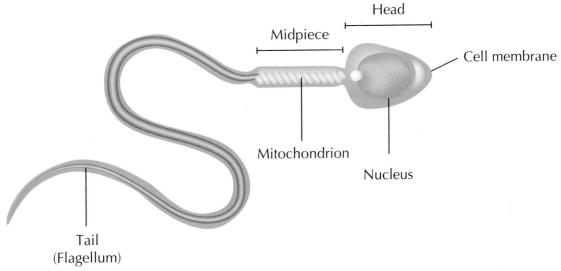

Fig 9.2 *Sperm cell*

The female gamete is the egg cell. This is much larger than the sperm cell, as it has a large store of food in its cytoplasm. The cytoplasm also contains mitochondria that are needed for mitosis (cell division) after fertilisation. The egg cell has a cell membrane surrounding the cell and contains a nucleus (fig 9.3).

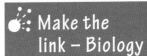

Make the link – Biology

Diploid cells have two sets of chromosomes (see page **92**).

Sperm and egg cells are ideally suited to their functions (see page **100**).

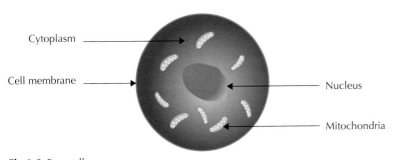

Fig 9.3 *Egg cell*

Reproductive structures in humans

The structure of the male reproductive system is shown in figure 9.4. Sperm are produced in a pair of organs called the **testes**. Sperm travel along the **sperm duct** towards the **penis**. As the sperm move along the sperm duct various glands add fluid to the sperm, forming semen. The penis introduces semen (including sperm) into the female's body to allow fertilisation to take place. Sperm leaves the penis through the urethra (fig 9.4).

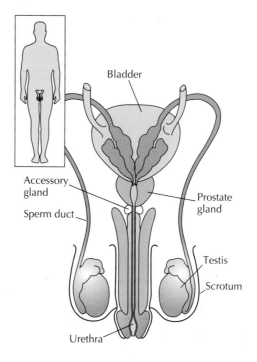

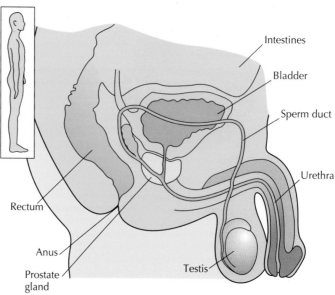

Fig 9.4 *The male reproductive system*

The structure of the female reproductive system is shown in figure 9.5. Eggs are produced in the **ovaries**. They travel along the **oviduct** towards the **uterus**. Sperm are deposited in the **vagina**. If an egg cell is fertilised by a sperm cell as it travels along the oviduct, it imbeds in the wall of the uterus and develops into a foetus (fig 9.5).

Hint

You need to be able to recognise the cells and organs involved in reproduction from diagrams.

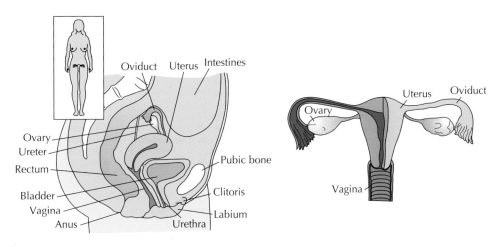

Fig 9.5 *The female reproductive system*

Biology in context

The menstrual cycle is a sequence of changes that takes place every month in a woman's body. The menstrual cycle is controlled by hormones. When levels of the hormone oestrogen increase, an egg is released from an ovary. This process is known as ovulation. Oestrogen also causes the lining of the uterus to thicken. Another hormone called progesterone helps the lining of the uterus to remain thickened. If the egg is not fertilised, it travels down the oviduct to the uterus. The thickened lining of the uterus comes away and leaves the body as a period. The period is made up of the unfertilised egg, blood and the lining of the uterus; it usually lasts around 3–7 days (fig 9.6).

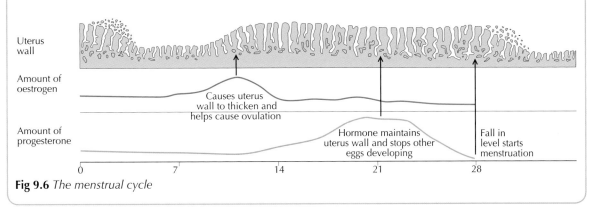

Fig 9.6 *The menstrual cycle*

GO! Activities

Activity 2.9.1 Working individually

1. Name the male and female gametes in animals.
2. Name the sites of gamete production in animals.
3. Explain why gametes must be haploid.
4. Describe how a sperm cell is suited to its function.
5. Describe how an egg cell is suited to its function.

Activity 2.9.2 Working in pairs

Sketch the male and female human reproductive system diagrams onto separate pieces of A5 paper. Label each structure, and below each label, describe its function.

I can:

- Explain that body cells are diploid (contain two sets of chromosomes), except gametes (sex cells), which are haploid (contain one set of chromosomes).

- State that in animals the male gamete is the sperm, which has a head and tail section and is produced in the testes.

- State that in animals the female gamete is the egg, which is larger than the sperm cell, has a large store of food in its cytoplasm and is produced in the ovaries.

- State that in animals the male reproductive system comprises the testes, which produce sperm, and the sperm duct, which allows sperm to pass to the penis. The penis then releases the sperm from the body.

- State that in animals the female reproductive system comprises the ovaries, which produce eggs, and the oviducts, which allow the egg to pass to the uterus, where a fertilised egg develops into a foetus.

Fertilisation in animals

Fig 9.7 *Like fish, frogs use external fertilisation*

Fertilisation is the fusion of the nuclei of the two haploid gametes to produce a diploid zygote, which divides to form an embryo. A zygote is a fertilised egg cell. Its nucleus contains two sets of chromosomes, one from each gamete. Fertilisation in animals can be divided into one of two types: external or internal.

Animals, such as fish, which live in water, often carry out external fertilisation. The female lays unfertilised eggs into the water. The male then releases sperm into the water. The sperm swim through the water and fertilise the eggs. This type of fertilisation is described as external because it takes place outside the organism's body.

Animals that live on land must use internal fertilisation. This is because sperm move towards egg cells by swimming, and there is no water in a land animal's immediate environment for the sperm to swim in. Sperm are deposited in the vagina along with watery semen. Sperm swim up through the uterus and along the oviducts. In animals, fertilisation takes place in the **oviduct**. The sperm nucleus fuses with the egg nucleus, and a zygote is formed. The zygote begins to divide, developing into an embryo as it travels along the oviduct to the uterus. It implants in the wall of the uterus and begins to develop (fig 9.9).

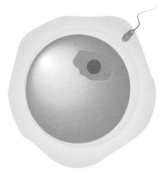

Fig 9.8 *Fertilisation occurs when the male gamete (sperm) nucleus fuses with the female gamete (egg) nucleus*

🌳 Biology in context

After fertilisation the zygote divides many times. When it has formed a ball of 64 cells it is known as an embryo. The embryo implants into the wall of the uterus and begins to develop. The embryo receives nutrients from its mother through the uterus lining. After 3 months it has developed into a foetus and receives nutrients from the placenta through the umbilical cord. In humans, the development of a foetus in the uterus takes 40 weeks. After this time, the foetus is fully developed and ready to be born.

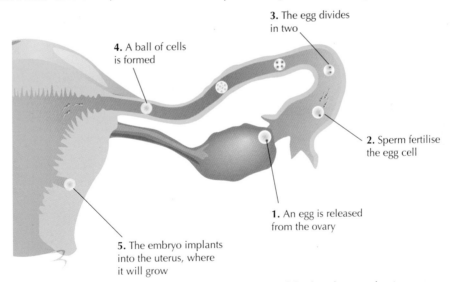

3. The egg divides in two

4. A ball of cells is formed

2. Sperm fertilise the egg cell

1. An egg is released from the ovary

5. The embryo implants into the uterus, where it will grow

Fig 9.9 *Internal fertilisation takes place in the oviduct of the female reproductive system*

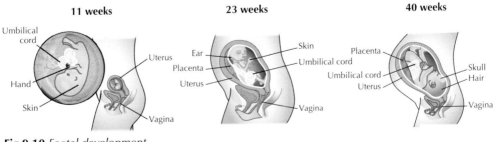

11 weeks   23 weeks   40 weeks

Umbilical cord

Uterus

Hand

Skin

Vagina

Ear

Placenta

Uterus

Skin

Umbilical cord

Vagina

Placenta

Umbilical cord

Uterus

Skin

Skull

Hair

Vagina

Fig 9.10 *Foetal development*

I can:

*   State that fertilisation in animals is the fusion of the nuclei of the two haploid gametes to produce a diploid zygote, which divides to form an embryo.

Reproduction in flowering plants

Like animals, plants can use sexual or asexual reproduction to produce new individuals. Many flowering plants use sexual reproduction. Typical flowering plants produce both male and female gametes in the same individual (fig 9.11).

The male parts of the flower are the **stamens** (fig 9.12). These are made up of an **anther** at the top and a stalk called a **filament**. The anther produces the male gamete, **pollen**. Pollen is usually transferred to other flowers to carry out fertilisation.

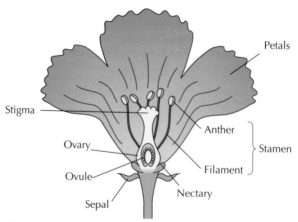

Fig 9.11 *The structure of a typical insect-pollinated flower*

The female part of the flower is made up of the **stigma** and the **ovary**. The stigma is where the pollen lands during pollination. The ovary produces the female gametes, the **ovules** (fig 9.13).

As well as male and female structures flowers may also have colourful petals. These attract insects, which can carry pollen from one plant to another. The sepal protects the petals before the flower opens. Flowers may also have nectaries, which produce a sweet liquid called nectar. This encourages insects to visit the flower. In various other species, pollen can be transferred by wind, water and other animals such as birds and bats.

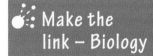

Make the link – Biology

Pollination is the process where pollen is transferred from one flower to another, by the wind or using insects (see page **133**).

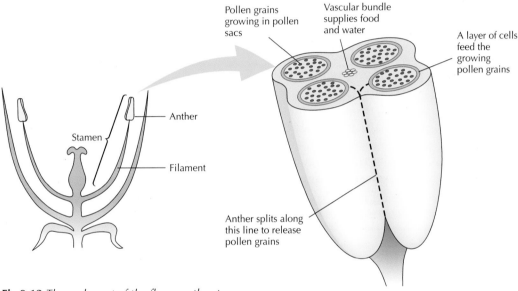

Fig 9.12 *The male part of the flower – the stamen*

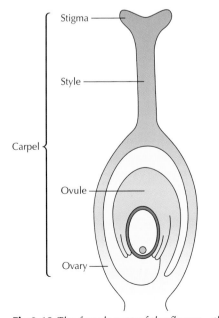

Make the link – Social Sciences

Many plants rely on insects for pollination and therefore reproduction of their species. Climate change may affect insect populations, which could have knock-on effects on world food production.

Fig 9.13 *The female part of the flower – the stigma and ovary*

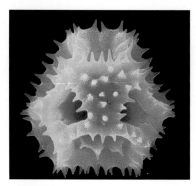

Fig 9.14 *A pollen grain from a dandelion*

⚕ Biology in context

Some people are allergic to pollen (fig 9.14) and experience an allergic reaction when it enters their eyes or breathing system. Hay fever is caused by wind-borne pollen. Different people are allergic to different types of pollen and can experience hay fever symptoms at different times of the year. In early spring, pollen from trees is the most common cause of hay fever, while in summer, grasses release pollen, which can affect hay fever sufferers.

Symptoms of hay fever include repeated sneezing, runny/itchy nose, watery eyes, itchy throat and a general feeling of being unwell. Very high pollen counts can also cause breathing to become wheezy. Treatments include antihistamine tablets, which help to reduce the allergic reaction. Anti-inflammatory nasal sprays and decongestant tablets may also be used.

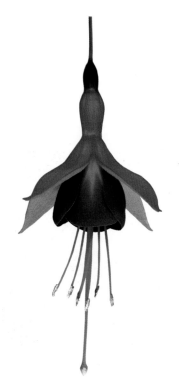

Fig 9.15 *A fuchsia flower*

GO! Activities

Activity 2.9.4 Working individually

Sketch the structure of a flower diagram in figure 9.13 onto a piece of A5 paper. Underneath each label, write down the function of the structure.

Activity 2.9.5 Working in pairs

Collect a flower (fuchsia flowers work well) and pull apart the different structures. Try to identify the following parts:

- Stamen (including anther)
- Stigma
- Ovary
- Petals
- Sepal

If you do not have a flower, try to identify these structures using the picture in figure 9.15.

Activity 2.9.6 Working in pairs

The shape, size and surface marking of a grain of pollen are unique to each species of plant. Pollen grains have a tough outer coat, which enables them to survive harsh conditions. Fossilised pollen has been found which was made millions of years ago!

Carry out some research to try and answer the following questions:

- What information does fossilised pollen provide us with?
- The pollen found in honey can be analysed; what will this tell us about the bees that made the honey?
- How can pollen be used to help solve crimes?

🌳 Biology in context

Before fertilisation can take place in flowering plants, pollination must occur. Pollination is the transfer of pollen from the anther to the stigma. There are two types of pollination. Self-pollination involves pollen moving from the anther to the stigma of the same plant. Cross-pollination involves the transfer of pollen between different plants.

Cross-pollination can be brought about by different methods. Some plants use the wind to transfer their pollen to other plants (fig 9.16). The anthers of wind-pollinated plants hang outside the flower. Their petals are small and dull because they do not need to attract insects. Their stigmas also hang outside the flower to catch pollen passing in the wind. Instead of using the wind to carry out pollination, many plants use insects. Plants that use insect pollination usually have brightly coloured petals and produce nectar and scents to attract insects (fig 9.17). When insects land on the flower, the pollen becomes attached to them, and they carry it to another plant where it can be brushed off.

Fig 9.16 *A wind-pollinated flower*

Fig 9.17 *An insect-pollinated flower*

Fertilisation in flowering plants

Fertilisation in flowering plants occurs when the haploid pollen nucleus fuses with the haploid ovule nucleus forming a diploid zygote. Eventually, the zygote will form a seed, and the wall of the ovary will develop into a fruit.

GO! Activity

Activity 2.9.7 Working individually
1. Name the male and female gametes in flowering plants.
2. Name the sites of gamete production in flowering plants.
3. Describe the process of fertilisation in flowering plants.

I can:

- State that in flowering plants the male gamete is pollen, which is produced by the anthers.

- State that in flowering plants the female gamete is the ovule, which is produced by the ovary.

- State that fertilisation in flowering plants is the fusion of the nuclei of the two haploid gametes to produce a diploid zygote.

10 Variation and inheritance

You should already know:

- The process of fertilisation involves the male and female sex cells fusing together to form a new organism.
- How to extract DNA from a tissue specimen.
- DNA controls our characteristics and contains the instructions for our cells to work properly.

Learning intentions

- Describe how sexual reproduction maintains variation.
- Give some examples of variation within species.
- Explain the difference between discrete and continuous variation.
- Explain the difference between polygenic and single gene traits and give examples of each.
- Identify examples of dominant and recessive traits.
- Explain what is meant by the term 'phenotype'.
- Give some examples of different phenotypes of a characteristic.
- Explain what is meant by the term 'genotype'.
- Assign a genotype to an individual.
- Explain what is meant by the terms 'homozygous' and 'heterozygous'.
- Explain what is meant by the terms 'gene' and 'allele'.
- Use the terms P, F_1 and F_2 in monohybrid crosses and family trees.
- Use Punnett squares to predict the inheritance of genes.
- Understand the use of family trees to investigate the inheritance of a characteristic.
- Explain why predicted ratios are not always achieved.

Variation

No two members of a **species** are exactly alike. Members of the same species are able to sexually reproduce to produce fertile offspring. During fertilisation, genes from each parent are passed onto their offspring. This means that sexual reproduction allows **genes** from **two parents** to be **combined**. This combining of genes results in differences between individuals of the same species. These differences are known as **variation**.

Variation can be seen in animals, for example tail length in red squirrels or antler height in red deer (fig 10.1). Variation also exists in plant species, for example petal colour in primrose plants and leaf size in oak trees.

Types of variation

There are two different types of variation: **discrete** variation and **continuous** variation. Characteristics showing discrete variation are inherited by combining alleles of a **single gene** only. An **allele** is a form of a gene. Discrete variation shows **single gene inheritance**.

Characteristics showing continuous variation are usually inherited when a characteristic is controlled by more than one gene. This means alleles for each gene are combined but the characteristic needs more than one gene to be expressed. Continuous variation shows **polygenic inheritance**. Polygenic inheritance means that more than one gene is involved.

- Discrete variation – controlled by a single gene. Characteristics **fall into distinct groups** and can be **measured**. For example, earwax in humans. There are only two types of earwax. You will either have dry earwax or wet earwax. There is nothing in between. Earlobes are another example of single gene inheritance. Humans can only have attached earlobes or unattached earlobes (fig 10.2).

- Continuous variation – controlled by more than one gene. Characteristics can have any value in a **range between the minimum and maximum value**. For example, height in humans. Humans can range in height from

Fig 10.1 *Amongst red deer there are many variations such as antler size or height*

🌳 Biology in context

In some rare cases, non-identical twins have been born with very different skin colours. Polygenic inheritance can explain this situation.

Skin colour is controlled by at least 20 genes. A mixed-race person will have some genes that contribute towards dark skin tone and some that contribute towards light skin tone. A mixed-race parent will pass on a combination of dark and light skin tone genes to their offspring. A twin with dark skin will have inherited more dark skin tone genes, and a twin with light skin will have inherited more light skin tone genes.

⚛ Make the link – Biology

Fertilisation is the process where gametes fuse, forming a zygote (see page **128**).

🔍 Hint

Remember that anything that needs to be measured, or counted, is most likely an example of continuous variation. E.g. number of freckles a person has. The height of an animal. The length of leaves in plants. These are all examples of continuous variation which are controlled by more than one gene.

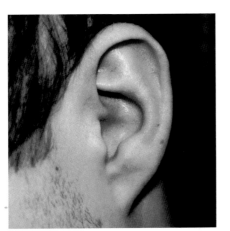
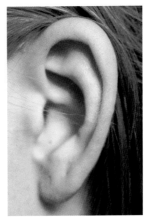

Fig 10.2 *Humans either have attached or unattached earlobes. This is an example of single gene inheritance*

🔍 **Hint**

Remember, '**poly**' means **lots**. So, polygenic inheritance is controlled by lots (more than one) of gene pairs.

Fig 10.3 *Skin colour varies greatly amongst the human population and is an example of polygenic inheritance*

around 60 cm to 250 cm once fully grown. Human skin colour is another example of polygenic inheritance since humans have many different colours of skin (fig 10.3). Figure 10.5 shows that human eye colour is an example of polygenic inheritance.

Polygenic versus single gene traits

Features such as flower colour in scarlet rosemallow and tongue-rolling ability in humans show discrete variation. These characteristics are controlled by the alleles of single genes. The ABO blood group system is controlled by one gene. This one gene gives rise to four different blood groups (A, B, AB and O), which show discrete variation.

Most characteristics shown by plants and animals are polygenic. This means they are controlled by many genes that act together. Polygenic characteristics show continuous variation. Beak depth in finches and height in humans are examples of polygenic characteristics.

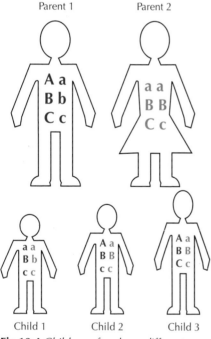

Fig 10.4 *Children often have different heights from their parents. This diagram shows how this is possible*

Imagine that just three genes control height in humans. Each gene comes in two different forms called alleles. One of the alleles can be given a capital letter symbol and codes for tallness. The other can be given a lower-case letter and codes for shortness. We each have two copies of every gene. Therefore a person who had the alleles AABBCC would be tall, and a person with alleles aabbcc would be small. There would be many people in between these extremes. These people would have different combinations of these alleles and therefore different heights (fig 10.4).

Fig 10.5 *Eye colour is controlled by at least three genes. These pictures show blue eyes of varying shades*

👁 Investigation

You can see variation in action in your own class. Carry out a survey to find out how many people in your class can roll their tongue. Copy and complete the table below.

	Can roll tongue	Cannot roll tongue
Number of pupils		

Draw a bar chart on a piece of squared paper to display your results. Explain whether tongue rolling is an example of discrete or continuous inheritance.

⠿ Make the link – Numeracy

Graph drawing is a skill used in maths as well as biology. You will use different types of graphs in biology, for example bar charts, line graphs and histograms.

GO! Activities

Activity 2.10.1 Working individually

1. Explain how variation arises in members of the same species.
2. Name the two different types of variation.
3. State the number of genes involved in controlling each type of variation.
4. Give a definition of the term single gene inheritance.
5. Give a definition of the term polygenic inheritance.

Activity 2.10.2 Working individually

Copy and complete the table below using the following list of characteristics.

- Flower colour in roses
- Shell diameter in limpets
- Height in emperor penguins
- Free or attached earlobe in humans (fig 10.6 and fig 10.7)
- Human hand span
- Gender in house sparrows

Discrete	Continuous

(Continued)

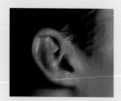

Fig 10.6 *An attached earlobe*

Fig 10.7 *A free earlobe*

Activity 2.10.3 Working individually

Bar charts are used to display information about discrete variation. Histograms are used to display information about continuous variation. An example of a histogram is shown below (fig 10.8).

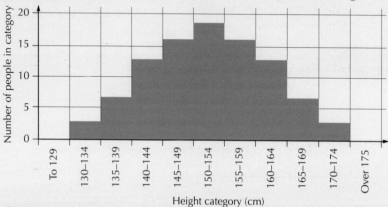

Fig 10.8 *A histogram*

The table below shows information about the hand span of a group of students.

Hand-span category (cm)	Number of students in category
15–16	2
17–18	6
19–20	4
21–22	1

To draw a histogram that displays this information, you will need to use an A5 piece of squared paper.

- Place each hand-span category along a scale on the bottom axis and label it.
- Make a suitable scale for the side axis and label it.
- Plot a bar for each hand-span category. The bars should be touching each other across the chart.

Activity 2.10.4 Working individually

1. In pea plants, the shape of the peas can be round or wrinkled.
   (a) Name the type of variation shown by this characteristic.
   (b) State whether this is likely to be a single gene or polygenic trait. Explain your answer.
2. In wheat, kernel colour can vary from dark red to white.
   (a) Name the type of variation shown by this characteristic.
   (b) State whether this is likely to be a single gene or polygenic trait. Explain your answer.

Activity 2.10.5 Working in pairs

Some conditions are caused by alleles of a single gene, whereas others are caused by the alleles of many genes. Carry out some research to discover which of the following conditions are caused by a single gene and which are polygenic.

Copy and complete the table below.

Cystic fibrosis	Asthma
Alzheimer's disease	Sickle-cell anaemia
Hypertension	Schizophrenia
Haemophilia	Diabetes

Single gene	Polygenic

I can:

- Give some examples of variation including long and short hair in cats, dry and wet earwax in humans, height in humans and leaf length in plants.

- State that sexual reproduction involves combining genes from two parents.

- State that combining genes from two parents contributes to variation within a species.

- State that discrete variation and continuous variation are the two types of variation.

- Explain that discrete variation is controlled by a single gene pair.

- Explain that polygenic inheritance is controlled by more than one pair of genes.

- Explain that in discrete variation measurements of characteristics fall into a distinct group.

- Explain that in continuous variation measurements of characteristics can have any value in a range.

Make the link – Biology

In Area of study 3, Chapter 19, you will learn more about adaptations. Adaptations arise due to mutations. Mutations are the only source of new alleles and are needed for variation to exist.

Hint

People often use the terms 'gene' and 'allele' as if they are the same thing. They are not. A gene is a portion of DNA that codes for a particular protein. An allele is a variation that exists for the gene.

Make the link – Biology

Diploid organisms have two matching sets of chromosomes (see page **92**).
Genes carry instructions for making proteins that control our characteristics (see page **44**).

Genetics terms

Genetics is the term used to describe the study of patterns of inheritance. To be able to discuss and understand genetics there are several terms that need to be used.

Gene versus allele

Chromosomes are structures found in the nucleus of cells. Each chromosome is made of DNA. A **gene** is a short portion of DNA. Each gene carries the code to make a specific protein (fig 10.9).

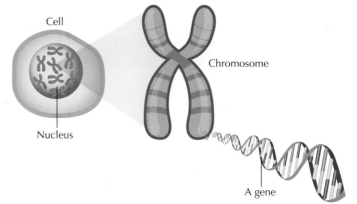

Fig 10.9 *A gene is a short section of DNA that codes for a specific protein*

The body contains genes for every protein that it makes. Humans receive one allele of each gene from their mother and one allele of each gene from their father. The code contained in the gene does not have to be identical.

Think of human hair colour. People have black, brown, blonde and red hair. One of the genes that controls hair colour has two alleles, blonde and brown. A person with one brown allele and one blonde allele will have brown hair (the dominant and recessive traits section explains why this happens). These variations for one gene are called **alleles** (fig 10.10).

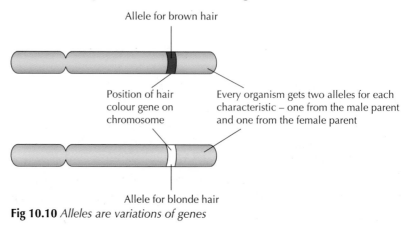

Fig 10.10 *Alleles are variations of genes*

Dominant and recessive traits

Diploid organisms have two copies of every gene. Alleles of genes can be described as **dominant** or **recessive**. For example, the gene for flower colour in pea plants has two different alleles. These are purple and white. If a plant with two purple alleles for the gene for flower colour is crossed with a plant with two white alleles for the gene for flower colour the offspring will all have purple flowers. This shows that the allele for purple flowers is dominant to the allele for white flowers (fig 10.11).

Purple parent has two purple alleles

×

White parent has two white alleles

All the offspring have one purple and one white allele

Fig 10.11 *The results of this cross show that purple is the dominant allele for flower colour in pea plants as all the offspring have purple flowers*

Dominant alleles always show up in the appearance of the organism, even if there is only one copy present.

Recessive alleles only show up in the appearance of an organism if they are paired with another copy of that recessive allele.

In humans, a person with one brown allele and one blonde allele for one of the genes that control hair colour will have brown hair. This shows that the brown hair allele is dominant, and the blonde hair allele is recessive (fig 10.10). In leopards, spotted coat is dominant to black coat (fig 10.12).

Allele 1	+	Allele 2	= appearance
Spotted coat	+	spotted coat	= spotted coat
Spotted coat	+	black coat	= spotted coat
Black coat	+	black coat	= black coat

Fig 10.12 *Spotted coat is dominant to black coat*

Phenotype and genotype

Phenotype describes the appearance of an organism. For example, in tulips, the phenotype for flower colour may be purple, red, yellow or pink (fig 10.13). In humans, the phenotype for earlobes may be unattached or attached (fig 10.2). Phenotype can sometimes be 'hidden', for example human blood groups can't be determined without medical tests.

An organism's **genotype** describes the set of alleles it possesses. When assigning an organism's genotype letters are used as symbols to represent the different alleles of a gene. Dominant alleles are given capital letters and recessive alleles lower-case letters. The same letter is used for the alleles of one characteristic.

In the fruit fly *Drosophila melanogaster*, one gene that controls body colour has two different alleles – yellow or ebony (black). The allele which results in yellow body colour is dominant to the allele that results in ebony body colour. A fly with two yellow body colour alleles has the genotype (**YY**). A fly with two ebony body colour alleles has the genotype (**yy**). A fly with the

🔍 Hint

Phenotype is how the genes that an organism possesses are expressed as **physical** characteristics. Phenotypes are usually shown as words. E.g. curly hair, freckles, white petals. **Genotypes** are the actual alleles for the gene that the organism possesses. These are always shown as a **pair of letters.**

Fig 10.14 *A fruit fly showing the dominant characteristic for body colour – yellow*

Fig 10.13 *Tulips have several different phenotypes for flower colour*

Fig 10.15 *Horses with a dominant A allele have black hairs clustered at their mane and legs. Horses with two recessive **a** alleles are black all over*

genotype (**Yy**) has one yellow allele and one ebony allele. Its genotype is (**Yy**) and its phenotype is yellow bodied (fig 10.14).

In horses, the agouti gene controls the distribution of black hairs in their coat. The dominant allele (**A**) causes black hairs to cluster at certain points (i.e. mane, tail, lower legs). This produces a bay-type horse. The recessive allele (**a**) causes black hairs to be spread evenly, producing a plain black horse (fig 10.15). The table below shows the different genotypes possible for this characteristic.

Genotype	AA	Aa	aa
What this tells us	This horse has two bay alleles	This horse has one bay and one black allele	This horse has two black alleles
Phenotype	Bay	Bay	Black

Hint

Be careful when deciding on a letter to use when writing out a genotype. It is easy to confuse some capital and lower-case letters such as U and u. Try to use a letter where the capital looks different from the lower case, for example G and g.

Biology in context

The fruit fly (*Drosophila melanogaster*) (fig 10.14) has been used in genetic studies for many years. *Drosophila* is an ideal organism to study genetic crosses as the individuals are small and easy to handle, they have a short generation time, and females produce many offspring at once. *Drosophila* is a model organism. By studying biological processes in *Drosophila*, we may gain some insight into how these processes take place in other organisms, such as ourselves. One very famous scientist to work with *Drosophila* was Thomas Hunt Morgan. Thanks to his work in the early 1900s, he put forward and proved several genetic theories. He was awarded the Nobel prize in physiology (medicine) in 1933 for his research into the role played by chromosomes in inheritance.

Homozygous and heterozygous

The words **homozygous** and **heterozygous** are used to describe the set of alleles an organism possesses for one gene – its genotype.

If the alleles are the **same** the organism is homozygous. If the alleles are **different** the organism is heterozygous.

For example, in pea plants one gene controls pea shape. The allele for round (**R**) is dominant to the allele for wrinkled (**r**). A plant with two round alleles (**RR**) is homozygous. A plant with two wrinkled alleles (**rr**) is also homozygous since the alleles it possesses are the same. A plant with one round and one wrinkled allele (**Rr**) is heterozygous. Figure 10.16 shows how different alleles influence whether an organism is describe as homozygous or heterozygous.

Hint

'Hetero' means different and 'homo' means the same. 'Zygous' refers to the zygote, which is the fertilised egg. So, heterozygous means that at fertilisation, the alleles the zygote received for one gene pair were different. Homozygous means that at fertilisation, the alleles the zygote received were the same.

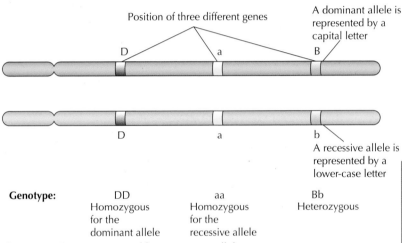

Fig 10.16 *Homozygous and heterozygous alleles*

P, F₁ and F₂

The above genetics terms are useful when interpreting family trees. A **family tree** shows the inheritance of a particular characteristic through several generations in one family. It is possible to identify **phenotypes** and **genotypes** using family trees.

Family trees use symbols to represent each individual. Males are represented using squares, and females are represented using circles. These symbols are coloured or left blank depending on if the individual shows the characteristic being investigated. A key is always provided to explain the family tree.

One gene controls the type of earwax humans produce. There are two possible phenotypes. These are wet earwax or dry earwax.

Make the link – Social subjects

In history, you may have learned about people from the past. Many people are interested in ancestry and tracing their family back for many generations. They may produce their family tree. A geneticist could use this information to trace the inheritance of genetic traits.

Hint

In family trees, males are always represented as a square and females are always represented as a circle.

Here is a family tree showing the inheritance of earwax type in one family (fig 10.17).

This family tree shows the following information:

- There are three generations in this family tree. There is the **parental generation P**, the **first generation** from the parental generation, the **F₁** individuals, and the **second generation** from the parental generation, **F₂**.

- The parental generation P (individuals A and B) had three children.

- The first generation children F₁ (C and G) each had two children. Their brother (E) had no children.

- Individuals F, G, I and M show the dry-earwax phenotype. All the other individuals have the wet-earwax phenotype.

- Individuals A and B both have wet earwax but had a child with dry earwax. Both parents must be heterozygous.

- This also shows that wet earwax is dominant to dry earwax.

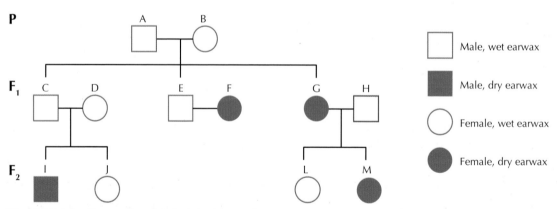

Fig 10.17 *A family tree showing the inheritance of earwax*

Biology in context

Queen Victoria reigned over the United Kingdom between 1837 and 1901. She is a direct ancestor of today's royal family (Queen Elizabeth II's great-great-grandmother). Queen Victoria was a carrier of haemophilia, a condition that reduces the blood's ability to clot.

Queen Victoria passed the haemophilia allele on to three of her nine children. Two were daughters, who became carriers of the condition and passed the haemophilia allele on to their children. One son received the haemophilia allele and was affected by the condition. He died at age 30 from bleeding after a fall (fig 10.18).

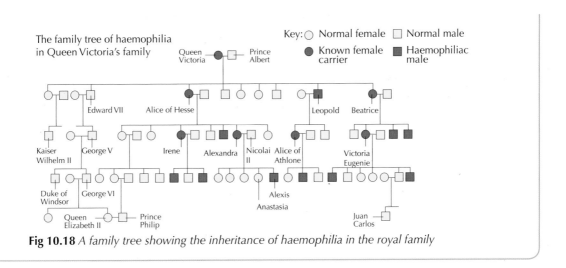

Fig 10.18 *A family tree showing the inheritance of haemophilia in the royal family*

GO! Activities

Activity 2.10.6 Working individually

1. Give a definition of a gene.
2. State the term used to describe different forms of a gene.
3. Give a definition of the term phenotype.
4. List the possible phenotypes for human eye colour.
5. Give a definition of the term genotype.
6. State the terms that can be used to describe different alleles.
7. Assign a genotype to an organism which has one dominant allele and one recessive allele for the gene for flower colour in pea plants. The flowers can be purple and white. Purple is the dominant allele.
8. Explain why the dominant allele will always appear in the phenotype of the organism.
9. Explain why a recessive allele never appears in the phenotype of an organism unless it possesses two recessive alleles.
10. Give a definition for the term heterozygous.
11. Give a definition for the term homozygous.
12. Name the type of diagram that can be used to show patterns of inheritance through generations of the same family.
13. Give the letter used to represent the parental generation in this type of diagram.
14. Give the letter used to represent the first generation in this type of diagram.
15. Give the letter used to represent the second generation in this type of diagram.

(Continued)

Activity 2.10.7 Working individually

1. Copy and complete the diagram below by filling in the missing word and diagram (fig 10.19).

Two _____ coat colour alleles   One black coat colour allele   Two black coat colour alleles
One white coat colour allele

Fig 10.19 *Coat colour in mice*

2. Using the information in the diagram, state which allele is dominant and which allele is recessive. Explain your answer.

Activity 2.10.8 Working individually

1. **(a)** Think of an organism (animal or plant).

   (b) Think of a variable characteristic that organism has.

   (c) Draw simple sketches of your organism to show different phenotypes of that characteristic and label each phenotype.

2. One gene controls the ability of the thumb to flex. It has two alleles, 'straight' or 'hitchhiker'. Straight is dominant to hitchhiker (fig 10.20). Use the letters H and h as symbols for the possible alleles for this gene.

   Copy and complete the following table.

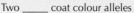

Fig 10.20 *A 'hitchhiker's' thumb*

Genotype	HH		hh
What this tells us		This person has one straight and one hitchhiker thumb allele	
Phenotype			

Activity 2.10.9 Working in pairs

The positions of four different genes are shown on figure 10.21 below. Copy and complete the diagram by:

- Filling in the genotype for each gene
- Stating whether this individual is homozygous or heterozygous for each gene.

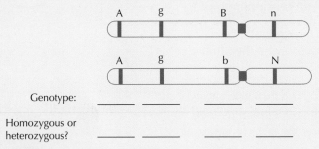

Genotype: ___ ___ ___ ___ ___

Homozygous or
heterozygous? ___ ___ ___ ___

Fig 10.21 *Three different genes*

Activity 2.10.10 Working in pairs

The family tree below shows the inheritance of freckles. Use the family tree to answer the questions that follow.

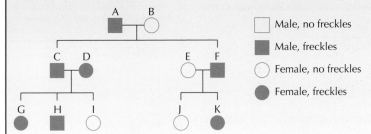

☐ Male, no freckles

■ Male, freckles

○ Female, no freckles

● Female, freckles

1. How many generations are shown in this pedigree chart?
2. How many individuals in this family have no freckles?
3. How many children did couple C and D have?
4. What are the phenotypes of individuals G, H and I?
5. Assuming individual A is heterozygous, which characteristic is dominant: freckles or no freckles?
6. Copy the family tree and assign a genotype to each individual.
7. Are there any individuals for whom you cannot be certain of their genotype? If so why?

I can:

- State that a gene is a portion of DNA that carries the code for a specific protein.

- State that an allele is a form of a gene.

- State that phenotype is a term used to describe the appearance of an organism.

- Give examples of phenotypes of one characteristic. For example, eye colour in humans (different phenotypes may be blue, brown or green).

- State that genotype is a term used to describe the alleles of the gene that an organism possesses.

- Assign a genotype to an organism based on the knowledge that dominant characteristics are given capital letters, and recessive characteristics are given lower-case letters.

- State that alleles can be dominant or recessive.

- Explain that the dominant allele will always appear in the phenotype of the organism.

- Explain that a recessive allele never appears in the phenotype of an organism unless it possesses two recessive alleles.

- State that an organism is heterozygous if the alleles that it possesses for one gene are different. State that homozygous organisms have both alleles the same.

- State that an organism can be homozygous if it has two identical alleles which are dominant for the trait or if its two identical alleles are recessive for the trait.

- State that a family tree is a diagram that allows the pattern of inheritance from one generation to the next to be determined.

- State that the parental generation in a family tree is represented by the letter P.

- State that the first generation of offspring in a family tree are represented by the symbol F_1.

- State that the second generation of offspring in a family tree are represented by the symbol F_2.

- Identify phenotypes and genotypes of individuals in a family tree.

Monohybrid cross

The Austrian monk, Gregor Mendel (fig 10.22), was the first person to study patterns of inheritance. He studied pea plants. He noticed that pea plants showed different phenotypes for the same gene (fig 10.23). He was interested to know why this happened and was able to determine how characteristics are passed on from parents to offspring using his experiments.

The inheritance of a characteristic which is controlled by one gene can be studied using a breeding experiment called a **monohybrid cross**. Monohybrid crosses follow the pattern of inheritance from the **parental generation** through to the F_2 **generation**. The original organisms used in a monohybrid cross are called the parental generation (**P**). These parents should be homozygous, but they should have a different phenotype. The offspring they produce are called the first generation (F_1), and the offspring these individuals produce are called the second generation (F_2).

Gametes are haploid. This means they have only one allele of each gene. To determine the genotype of the offspring in a monohybrid cross the genotype of the gametes produced by the parental generation must be determined. When the gametes fuse, a new individual is formed with two alleles of the gene that is being studied.

One example of a monohybrid cross is shown in figure 10.24.

In this cross, all the F_1 generation have purple flowers. This is because purple is the dominant allele for the gene for flower

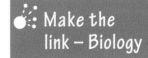

Make the link – Biology

In Area of study 3, Chapter 19, you will learn that some new alleles, created by mutation, give a selective advantage to the organism. The new allele makes the organism better adapted to its environment. It therefore survives and is able to pass on its favourable allele to its offspring.

Not all new alleles are favourable. Sickle cell anaemia is a condition in humans caused by offspring inheriting two sickle cell alleles.

Can you work out if sickle cell anaemia is caused by a dominant or recessive trait?

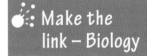

Make the link – Biology

Gametes (sex cells) are haploid and fuse together during fertilisation (see page **128**).

Biology in context

Gregor Mendel was an Austrian monk, who carried out experiments investigating inheritance in the mid-1800s. Thanks to his work with pea plants, he was able to

determine how characteristics are passed on from parents to offspring. Mendel came up with the idea that pea plants had two 'factors' for each characteristic, one inherited from each parent. We now know that these 'factors' are genes. Mendel also noted that some traits that did not show up in an individual pea plant could be passed on to the next generation. In other words he was noticing the effects of dominant and recessive alleles.

Mendel's theories at this time were very advanced, and many scientists misunderstood or disagreed with his ideas. It wasn't until the early 1900s that the importance of Mendel's investigation was recognised. Mendel died in 1884, having never received recognition for his pioneering work (fig 10.22).

Fig 10.22 *Mendel is known as the 'Father of genetics'*

colour. The genotypes of these flowers are the same – Gg. This is because each flower received one dominant allele from the purple flowered parent and one recessive allele from the white flowered parent.

The F_1 generation are the parents of the F_2 generation. As the F_1 parents are heterozygous a **Punnett square** is used to work out the genotypes of the **F_2** generation. One parent's gametes are placed along the top of the Punnett square and the other parent's gametes along the side. The genotype of the offspring can then be filled in.

In the F_2 generation there are some purple flowers and some white flowers. There are more purple flowers as there is a higher chance of a dominant purple allele being present in the zygote compared to the chance of two white alleles being present in the zygote. The **expected ratio** of phenotypes showing the dominant phenotype to those showing the recessive phenotype is **3:1**.

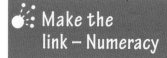

Make the link – Numeracy

In maths, you might have learned about simple whole number ratios. Biologists use simple whole number ratios also. If you simplify a ratio and your answer contains a decimal it would be a really good idea to check your working!

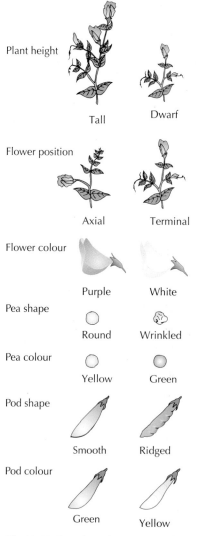

Plant height — Tall / Dwarf

Flower position — Axial / Terminal

Flower colour — Purple / White

Pea shape — Round / Wrinkled

Pea colour — Yellow / Green

Pod shape — Smooth / Ridged

Pod colour — Green / Yellow

Fig 10.23 *Pea plants have many characteristics that can vary between individual plants. Dominant characteristics are shown on the left, and recessive characteristics are shown on the right*

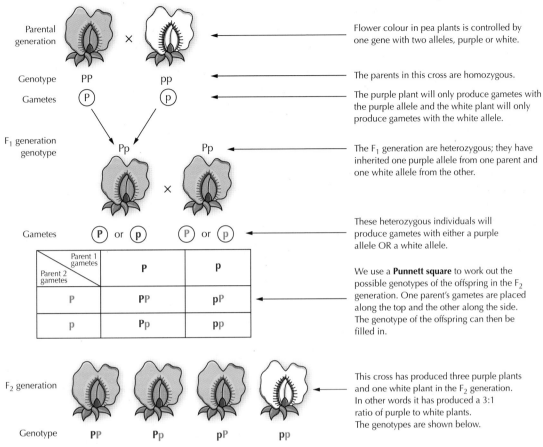

Fig 10.24 *A monohybrid cross*

The following annotations appear alongside the diagram:

- Flower colour in pea plants is controlled by one gene with two alleles, purple or white.
- The parents in this cross are homozygous.
- The purple plant will only produce gametes with the purple allele and the white plant will only produce gametes with the white allele.
- The F₁ generation are heterozygous; they have inherited one purple allele from one parent and one white allele from the other.
- These heterozygous individuals will produce gametes with either a purple allele OR a white allele.
- We use a **Punnett square** to work out the possible genotypes of the offspring in the F₂ generation. One parent's gametes are placed along the top and the other along the side. The genotype of the offspring can then be filled in.
- This cross has produced three purple plants and one white plant in the F₂ generation. In other words it has produced a 3:1 ratio of purple to white plants. The genotypes are shown below.

Predicted phenotype ratios

The **predicted ratio** of a monohybrid cross, where the parental generation are homozygous but show different phenotypes, is 3:1. The dominant phenotype is expected to appear three times more in the **offspring** than the recessive phenotype.

This predicted ratio is **not always achieved** when the cross is carried out. The reasons for this include:

- The sample size – there were not enough offspring produced to allow statistical analysis to achieve a ratio of three to one.

- Chance – fertilisation is a random process. The following investigation demonstrates the role played by chance on the fertilisation of gametes.

◉ Investigation

One of the features that Gregor Mendel studied in pea plants was pod colour. There are two phenotypes for pod colour: green (dominant) and yellow (recessive) (fig 10.23).

We can investigate the inheritance of this characteristic using different-coloured beads to represent each allele. Green beads represent the dominant green allele, and yellow beads represent the recessive yellow allele. Follow the instructions below to perform a cross between a homozygous green pod plant and a homozygous yellow pod plant.

- Place 20 green beads into one beaker. This represents the gametes produced by the green pod parent.
- Place 20 yellow beads into another beaker. This represents the gametes produced by the yellow pod parent.
- To model the offspring produced by these parents, take one gamete (bead) from one pot and pair it with one gamete (bead) from the other pot.
- Write down the genotype and phenotype of the offspring, using 'G' for green alleles and 'g' for yellow alleles.

What do you notice about the genotype and phenotype of the offspring in this cross? Can you explain this result?

Planning and designing an investigation

Describe and explain how you could adapt this investigation to show a cross between two heterozygous individuals. (Hint: you will still need 20 beads in each beaker but think about their colour.)

A group of students performed an investigation like this using heterozygous parents. They used the beads to produce 20 offspring.

Making a prediction

What do you think their results would be? How many green-pod and how many yellow-pod offspring would you expect there to be?

Results

The results from their investigation are shown below.

Number of green/green pairs in offspring	4
Number of green/yellow pairs in offspring	13
Number of yellow/yellow pairs in offspring	3

Draw a Punnett square for this cross. (Remember the parents were both heterozygous (Gg).)

Conclusion

What phenotype ratio would you predict in the offspring?

How do the actual results differ from the expected results?

Evaluation

Explain why the actual results may have differed from the expected results.

How could you adapt the investigation to help increase the reliability the results?

GO! **Activities**

Activity 2.10.11 Working individually

1. Name the type of cross which follows the pattern of inheritance from the parental generation through to the F_2 generation.
2. Explain why a Punnett square is used in this cross.
3. State the F_2 predicted ratio for this type of cross.
4. Explain why achieved ratios may not be the same as the predicted ratio.

Activity 2.10.12 Working in pairs

Pea shape can differ between individual pea plants. There are two phenotypes for this characteristic: round (dominant) and wrinkled (recessive). Two individual pea plants were crossed. One was homozygous for round pea shape and one was homozygous for wrinkled pea shape. The F_2 generation contained 240 pea plants.

1. Using the symbol 'R' for round and 'r' for wrinkled, write out the genotype of each parent.
2. Write out the gametes each parent could produce underneath.
3. What would be the genotype and phenotype of the offspring (F_1 generation) produced in this cross?

Two of the F_1 generation were then crossed.

1. Write out the genotype of these individuals.
2. Underneath write out the gametes each parent could produce.
3. Draw a Punnett square and fill it in to determine the possible genotypes of the offspring (F_2 generation).
4. How many round pea plants would you expect to see in the F_2 generation?
5. How many yellow pea plants would you expect to see in the F_2 generation?

Activity 2.10.13 Working individually

In pea plants the gene for plant height has two alleles. These are tall and dwarf. A dwarf plant was crossed with a tall plant. Both plants were homozygous. Tall is the dominant allele.

Use figure 10.24 as a template to show the inheritance of height by the F_2 generation.

Activity 2.10.14 Working individually

• The cross described in activity 2.10.13 was carried out by students and the results they obtained are shown in the table below.

Height of plant	Number of plants in F_1	Number of plants in F_2
Tall	65	240
Dwarf	0	60

• State the predicted ratio of the F_2 generation.
• Work out the actual ratio of the F_2 generation.
• Explain why the actual ratio obtained is different from the predicted ratio.
• State the percentage of dwarf plants in the F_2 generation.

I can:

- Carry out monohybrid crosses.

- Use Punnett squares to explain patterns of inheritance.

- State that predicted ratios are not always achieved when carrying out monohybrid crosses.

- Explain that observed ratios are influenced by the sample size and fertilisation involving the element of chance.

11 Transport systems – Plants

You should already know:

- Plants play a vital role in sustaining life on Earth.
- Plants use light energy to convert carbon dioxide and water into oxygen and sugar by a process known as photosynthesis.

Learning intentions

- Name plant organs and describe their function.
- Name the different types of cell found in a leaf.
- Describe the role of stomata.
- Explain the need for transport systems in plants and explain the role played by water.
- Name the parts of plants involved in transporting water.
- Give details of the structure and function of xylem.
- Explain what is meant by the term 'transpiration'.
- List factors which affect the rate of transpiration.
- Give details of the flow of water through a plant.
- Describe the structure and function of phloem cells.

⁂ Make the link – Biology

In Area of study 2, Chapter 7 you learned about cells, tissues and organs. Look back at this chapter to remind yourself what these terms mean.

Plant organs

Plants are living organisms which contain specialised cells. Most plants are multicellular but single celled plants also exist. Multicellular plants contain **organs**. The cells in these organs are adapted for specific functions. The organs of plants include **roots**, **stems** and **leaves**.

Plant roots allow the plant to take up water and nutrients from the soil. They also keep the plant upright in the soil. The stem connects the roots to the leaves. It allows transport of water, sugar and nutrients to different parts of the plant. The leaves of the plant allow photosynthesis to occur. Photosynthesis is the process that plants use to produce food.

Some types of plants have flowers. The flowers are involved in reproduction. A fertilised flower will eventually form a fruit (fig 11.1).

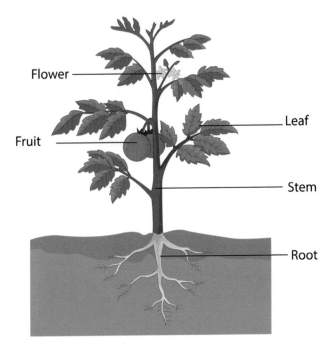

Fig 11.1 *The plant organs root, stem and leaves can be observed in a tomato plant*

Leaf structure

Leaves are the plant organs which are the main site of photosynthesis (fig 11.2). They contain different types of specialised structures to allow them to photosynthesise efficiently (fig 11.3). The functions of these structures are found in the table below.

Name of structure	Function
Stomata	Tiny pores that allow exchange of gases with the air
Upper epidermis	A layer of colourless cells that allows light to pass through to the next layer
Palisade mesophyll	The main site of photosynthesis; these cells contain lots of chloroplasts
Spongy mesophyll	Cells surrounded by air spaces to allow gases to circulate in the leaf and reach leaf cells
Leaf vein	Contains **xylem** and **phloem**, which allow water and sugar to be transported into and out of the leaf
Lower epidermis	Contains guard cells, which form tiny pores called stomata which allow gases to pass in and out the leaf
Guard cells	Control whether the stomata are open or closed

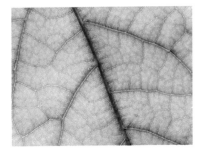

Fig 11.2 *Leaves are well adapted to carrying out photosynthesis. They have a large surface area to absorb maximum light and are thin to allow light to reach all the cells and gases to exchange efficiently*

Hint

Make sure that you can name, and identify, each type of cell shown in figure 11.3. This is an area frequently asked about in exams. There are lots of new words to remember!

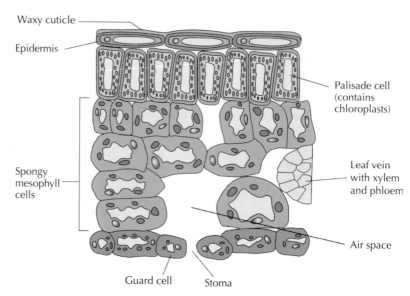

Waxy cuticle

Epidermis

Palisade cell (contains chloroplasts)

Spongy mesophyll cells

Leaf vein with xylem and phloem

Air space

Guard cell

Stoma

Fig 11.3 *The cells of a leaf*

Stomata

Stomata (singular: stoma) are pores found on the surfaces of leaves. Stomata allow gases to move into and out of the leaf. Carbon dioxide reaches photosynthesising cells in the leaf by entering through stomata, and oxygen exits in this way. The stomata also allow water vapour to leave the plant by evaporation.

Two guard cells surround each stoma (fig 11.4).

Most of the stomata are found on the lower surface of the leaf (fig 11.5). This reduces the volume of water lost by the plant since the stomata are protected from drying winds and sunlight. Stomata are found on the upper surfaces of aquatic plants to allow gas exchange to occur. If the stomata were on the lower surfaces water could block the stomata (fig 11.6).

Fig 11.4 *Guard cells surround the stoma*

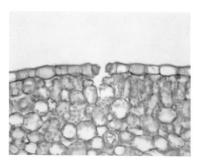

Fig 11.6 *Aquatic plants have stomata on their upper leaf surfaces to allow gas exchange*

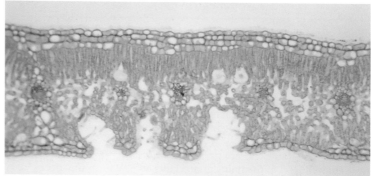

Fig 11.5 *A cross section through a leaf*

🔵 Activities

Activity 2.11.1 Working individually

1. List three organs which are found in plants and for each state its function.

2. Name the layer of colourless cells in a leaf that allows light to pass to the cells below.

3. Explain why spongy mesophyll cells are surrounded by air spaces.

4. Explain how the structure of a palisade mesophyll cell is related to its function.

5. Name the pores found on the surface of a leaf. Explain why most of these pores are found on the lower surfaces of leaves.

Activity 2.11.2 Working individually

Collect a privet leaf and some clear nail polish. Paint a layer of nail polish onto the lower surface of the leaf and allow it to dry. Peel off the nail polish, place it on a slide and look at it under the microscope.

How could this experiment be used to estimate the density of stomata on the surface of a leaf?

Activity 2.11.3 Working in pairs

Discuss the following questions with a partner and note down your ideas.

1. Water lilies have leaves that float on the surface of the water in which they grow. Where do you think their stomata are found? Explain your answer.

2. Some plants have vertical leaves which are equally exposed to the sun. How do you think this would affect the number of stomata on their upper and lower surfaces?

3. Some plants have very few, small stomata, which reduces the volume of water they lose. What type of habitat might these plants live in?

Activity 2.11.4 Working in groups

When you think of a plant you probably think of the parts that grow above the ground. For that reason, you will be more familiar with the plant organs stems and leaves than you are with the plant organ roots. You may wish to study plant roots in more detail. To do this you should collect a variety of seeds – cress, mustard and broad beans work well. Place some cotton wool in the bottom of a Petri dish and add enough water to soak it. Place the seeds on top of the cotton

(Continued)

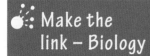

Make the link – Biology

When a leaf cell gains water, it becomes turgid. When it loses water, it becomes plasmolysed (see page **34**).

wool and leave for around five days to grow – broad beans will require more time than cress or mustard seeds.

When leaves can be seen the roots should be well established. Gently remove the germinated seed from the cotton wool and examine the roots with a hand lens. You should be able to see root hairs. Root hairs increase the surface area of the root so that the seedling can take up the maximum volume of water (figs 11.8 and 11.9).

I can:

- State that plants contain three organs, which are roots, stems and leaves.

- Name and identify from a diagram the structures found in leaves as upper and lower epidermis, palisade and spongy mesophyll, guard cells, stomata and the leaf vein which contains xylem and phloem.

Make the link – Biology

Photosynthesis requires water and carbon dioxide to produce sugar and oxygen (see page **274**).

Biology in context

Water is very important to plants. Water is needed by plants for:

- Photosynthesis – water is a raw material in the process of photosynthesis. It is combined with carbon dioxide to make sugar for the plant to use in respiration.
- Absorbing minerals – minerals are dissolved in soil water. When plants take in water they also take in the dissolved minerals they require. These minerals are transported to where they are needed in the water transport system.
- Maintaining turgidity – plant cells which lack water become plasmolysed. Plasmolysed cells cannot carry out their function properly.
- Preventing damage due to heat – water evaporates from the stomata of plants. The evaporation of water cools the plant and prevents damage to the leaves.

Water transport in plants

Plants require a **transport system** to move **water** around the plant. This is because water is a vitally important substance for plants. Plants must have a transport system to allow water to move from where it is absorbed to where it is needed (fig 11.7).

The **root hair** cells create a large surface area for the absorption of **water** and dissolved **minerals** by the roots (fig 11.8 and fig 11.9).

Water enters plants by osmosis through root hair cells. It then moves from cell to cell (also by osmosis) until it reaches a **xylem**

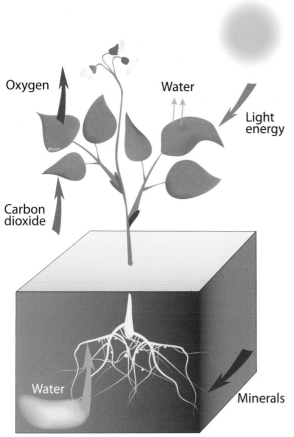

Oxygen

Water

Light energy

Carbon dioxide

Water

Minerals

Fig 11.7 *Plants need water to carry out photosynthesis. Water is transported from the roots, where it is absorbed, to the leaves, where photosynthesis takes place*

⚗️ Make the link – Technologies

Xylem tissue gives timber its grain patterns and affects its strength. It is important to choose the correct type of wood when building furniture.

🔍 Hint

Try to remember **x-wam**, **x**ylem transports **w**ater **a**nd **m**inerals.

Fig 11.8 *The microscopic root hairs can be seen due to high magnification. Root hairs increase the surface area for absorption of water*

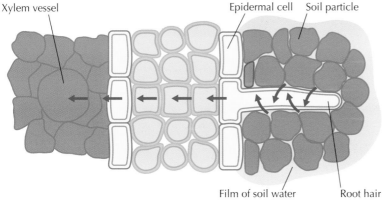

Xylem vessel

Epidermal cell   Soil particle

Film of soil water   Root hair

Fig 11.9 *Root hair cells have extensions that project into the soil. This increases the surface area for the absorption*

vessel. Xylem vessels are tubes of **dead**, hollow cells that run through the entire length of a plant (fig 11.10).

Water is transported up through the stem in the xylem vessels from the roots where it is absorbed to the leaves where it is used. When water reaches the leaves, it moves from the xylem

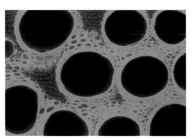

Fig 11.10 *Xylem vessels are dead as they do not contain a nucleus. They are hollow and run the entire length of the plant*

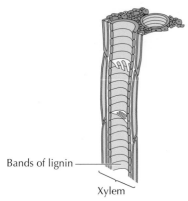

Bands of lignin ——

Xylem

Fig 11.11 *Xylem allows water and minerals to be transported from the roots to the leaves*

vessels into the mesophyll cells of the leaf by osmosis. The water then evaporates out of the mesophyll cells into the air spaces in the leaf. Finally, it leaves the leaf (fig 11.7) by diffusing out through the stomata (fig 11.4).

Structure of xylem vessels

Xylem is often strengthened by a substance called **lignin**. Lignin allows the plant to withstand **changes in pressure**. The pressure changes in the xylem vessels are caused by **water moving** through the plant. Containing lignin is an adaptation to windy conditions. It strengthens plant stems making them less likely to be damaged by strong winds. Lignin spans the length of the xylem vessels (fig 11.11) and takes the form of rings or spirals (fig 11.12).

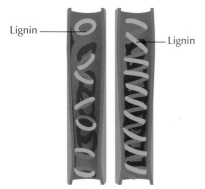

Lignin ——

—— Lignin

Fig 11.12 *Lignin can be found in rings or spirals*

Biology in context

As well as transporting water and minerals, xylem has a secondary function – support. In perennial plants, xylem develops in rings. These rings can clearly be seen in trees. Much of the wood of a tree is made up of xylem tissue. Each year a new ring of xylem develops on the outside of a tree trunk, and this allows us to estimate the age of a tree (fig 11.13).

Fig 11.13 *This image shows the rings on the trunk of a tree, which allow its age to be estimated*

GO! Activities

Activity 2.11.5 Working individually

1. State the parts of the plant that are involved in the transport of water.
2. State the exact location where water enters a plant.
3. Name another substance that will enter the plant in the water.
4. Explain how the plant obtains the maximum surface area to absorb water.
5. Name the type of vessels that transport water throughout the plant.
6. State whether these vessels are alive or dead. Explain your answer.
7. Name the substance that strengthens xylem vessels.
8. Describe two ways this substance can be found in xylem vessels.
9. Explain the need for this substance in xylem vessels.

Activity 2.11.6 Working in pairs

Place a cut stem of celery (with leaves on is best) in a beaker containing dyed water. Leave for 24 hours and then cut it in half across the stem. Have a look at the cut surface that was dipped into the dye (fig 11.14).

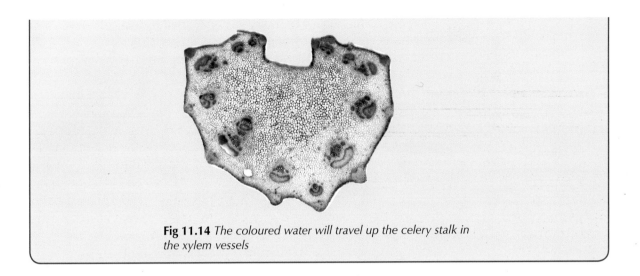

Fig 11.14 *The coloured water will travel up the celery stalk in the xylem vessels*

I can:

- State that plants require transport systems to move substances such as water and minerals from one place to another.

- State that plants absorb water and dissolved minerals through root hairs.

- State that water and minerals are transported from the root hairs to the xylem vessels.

- State that xylem vessels are dead.

- State that xylem vessels are strengthened by lignin.

- Explain that lignin allows the xylem cells to withstand pressure changes that occur as water moves through the plant.

Transpiration

Thinking about experiments 3

Water molecules are small and move from cell to cell by osmosis. In a leaf, water evaporates from the xylem into the air spaces between the spongy mesophyll cells, creating water vapour. If the stomata are open, the water vapour will diffuse out of the leaf into the air (fig 11.15). This water loss is known as **transpiration**.

The stomata must be open to allow gas exchange. However, this causes a high level of transpiration. Plants must perform a balancing act to obtain gases for photosynthesis and respiration but ensure they do not lose too much water. One way to achieve this balance is for the guard cells to close the stomata during the night to save water when the plant cannot photosynthesise (fig 11.16). Water moving through the plant and evaporating through the stomata is called transpiration.

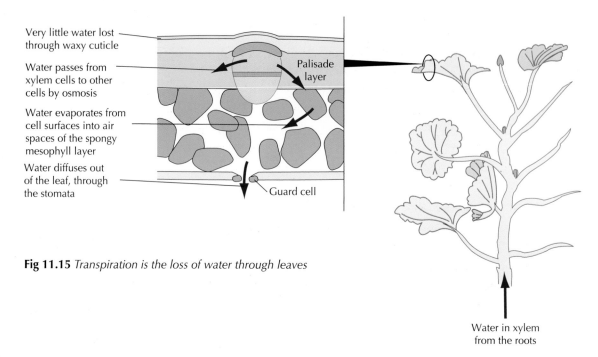

Very little water lost through waxy cuticle

Water passes from xylem cells to other cells by osmosis

Palisade layer

Water evaporates from cell surfaces into air spaces of the spongy mesophyll layer

Water diffuses out of the leaf, through the stomata

Guard cell

Water in xylem from the roots

Fig 11.15 *Transpiration is the loss of water through leaves*

Make the link – Biology

Osmosis is the movement of water molecules from an area of high water concentration to an area of low water concentration through a selectively permeable membrane (see page **33**).

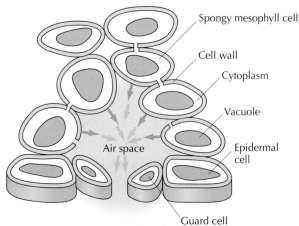

Spongy mesophyll cell

Cell wall

Cytoplasm

Vacuole

Air space

Epidermal cell

Guard cell

Fig 11.16 *Water evaporates into the air spaces in the leaves and moves out through the stomata*

Hint

Four external factors affect the rate of transpiration. Make sure you can list wind speed, humidity, temperature and surface area of the leaf as these factors.

Biology in context

Plants which live in very hot places would die if they didn't possess adaptations to allow them to survive. Very hot places tend to be very dry places. This means that plants must not lose water by transpiration or else they will die. The problem is that these plants will also die if they do not open their stomata to obtain carbon dioxide needed for photosynthesis. It seems a catch 22 situation. However, these plants have adapted a way to open their stomata at night when it is cooler. They store carbon dioxide in their leaves until morning and then they begin to photosynthesise.

Factors affecting the rate of transpiration

The **rate of transpiration** is affected by several **external factors**. These external factors can increase or decrease the rate of transpiration. Transpiration can be affected by wind speed, humidity, temperature and surface area of the leaf.

Wind speed – the higher the wind speed the higher the rate of transpiration. Wind removes water vapour from the stomata. If this water vapour is constantly removed the plant will need to replace it.

Humidity – the higher the humidity the lower the rate of transpiration. High humidity means that there is a lot of water vapour already in the air. The plant will not lose as much water.

Temperature – the higher the temperature the higher the rate of transpiration. High temperatures cause water to evaporate and the plant needs to transpire to replace this water.

Surface area of leaf – the larger the surface area of the leaf the higher the rate of transpiration. This is because the plant has a larger area over which to lose water. To compare transpiration rates between different plants, the same surface area should be compared. This is done by measuring the transpiration rate per square unit of surface area.

⚬ Make the link – Biology

Transpiration rate can be measured using a potometer. There are two types of potometer – a weight potometer and a bubble potometer. The investigation below shows how a simple weight potometer can be used to investigate the effect temperature has on the rate of transpiration. You are also asked to think about how this experimental set up could be altered to investigate the effect light intensity has on transpiration.

Thinking about experiments 3, page **353**, explains the use of potometers to measure transpiration rates in more detail.

👁 Investigation

The experiment shown can be set up to investigate the effect of temperature on the rate of transpiration (fig 11.17).

Planning and designing an investigation

A layer of oil is placed over the water surface. Explain why this is necessary.

Design an experiment that would allow you to investigate the effect of temperature on the rate of transpiration.

What factors would need to be kept the same in this investigation?

Making a prediction

How do you think temperature will affect transpiration? Write down your hypothesis.

Results

Here are some sample results from an experiment to investigate the effect of temperature on the rate of transpiration.

Temperature (°C)	Water loss (g)
5	0.45
20	2.25
30	3.30
40	5.40

Fig 11.17 *This potometer is weighed at the start of the experiment and again after 24 hours*

Copy and complete the following table to show the rate of transpiration in grams of water lost per hour (remember the experiment was set up for 24 hours).

Temperature (°C)	Rate of transpiration (g per hour)
5	
20	
30	
40	

On a piece of A5 graph paper, draw a line graph to display these results.

Conclusion

State the relationship between temperature and water loss.

State a conclusion relevant to the aim of this investigation.

Evaluation

How could the reliability of this experiment be improved?

 Activities

Activity 2.11.7 Working individually

1. Describe the process of transpiration.
2. State where and by what process water enters plants.
3. Explain how water moves through plants till it reaches the leaves.
4. Name the vessels that transport water upwards through a plant.
5. State the final fate of water travelling through a plant.
6. List external factors that affect the rate of transpiration.

Activity 2.11.8 Working individually

- Look at figures 11.15 and 11.16. Use these to help you write a short note explaining transpiration. Include the following words in your explanation:

  evaporation   osmosis   diffusion   transpiration

Activity 2.11.9 Working in pairs

Plants can be classified into different groups depending on the adaptations they have to extreme conditions. Some plants are classified as xerophytes.

Carry out research on xerophytes. You should find out the adaptations that these plants have which allow them to live in dry conditions. Their adaptations allow these plants to transpire without losing too much water.

Include in your findings:

1. Two different types of ecosystem that these plants live in.
2. Reasons why there is a shortage of water available to the plants.
3. Adaptations to their:
   * leaves
   * stems
   * roots
   * stomata

   that allow these plants to survive in the dry conditions.

You should present your findings as a poster which includes a labelled diagram showing the adaptations.

I can:

* Describe the process of transpiration as the movement of water through a plant until it evaporates through the stomata.
* State that water enters the plant root hairs by osmosis.
* Explain that water then travels from cell to cell by osmosis till it reaches the xylem vessels.
* State that water moves up the plant through xylem vessels.
* Explain that water eventually evaporates through the stomata.
* State that the rate of transpiration is affected by external factors.
* List wind speed, humidity, temperature and leaf surface area as external factors that affect the rate of transpiration.

Transport of sugar through plants

Sugar is transported around the plant in **phloem cells**. Phloem cells are linked together to form a continuous tissue that moves sugar to wherever it is needed (fig 11.18). Unlike xylem, phloem cells are **living**. Phloem cells are connected at **sieve plates** that allow substances to pass easily from cell to cell (fig 11.19). The **associated companion cells** control the opening and closing of the sieve plates.

Make the link – Biology

In Area of study 1, Chapter 4, you learned about enzymes. Can you explain why a potato stores sugar as starch? Can you also explain how this starch is used in respiration?

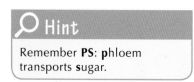

Hint

Remember **PS**: **p**hloem transports **s**ugar.

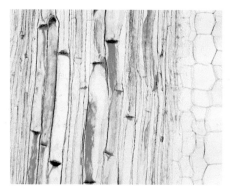

Fig 11.19 *Phloem tissue in the stem of a plant contains sieve plates and companion cells*

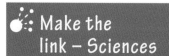

Make the link – Sciences

In physics you may have learned about gravity. Gravity is a force that pulls objects downwards. Companion cells ensure sugar is transported to where it is needed by the phloem tissue.

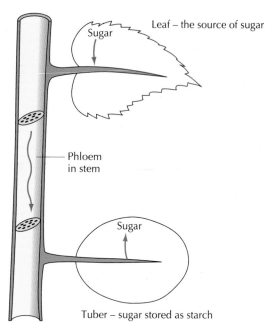

Fig 11.18 *Plants use phloem cells to transport sugar from one part of the plant to another. For example when forming a tuber (e.g. a potato) sugar is transported from the leaves to the growing tuber*

GO! Activities

Activity 2.11.10 Working individually

1. Name the substance which is transported in the phloem **and** describe the direction in which this substance is transported.
2. Name the two types of cells that are found in phloem.
3. State the function of each of these types of cells.
4. Explain whether phloem is considered to be dead or alive.

Activity 2.11.11 Working in pairs

- Potatoes are a store of sugar. Use the internet to research how potatoes are formed.

Activity 2.11.12 Working in groups

- Carry out an investigation into potatoes. You should take a small piece of potato and test it for:
  - sugar using Benedict's reagent and
  - starch using iodine.
- It is anticipated that your results will show that there is starch present in potatoes. Use your biological knowledge to explain this result.

Activity 2.11.13 Working individually

Copy and complete the table below comparing xylem and phloem.

	Xylem	Phloem
Material transported	Transports water and minerals	
Composition		Made up of living cells
Direction of transport	Transports water and minerals upwards only	
Special features	Contains lignin	

Activity 2.11.14 Working in pairs

Make an A4 information sheet about transport systems found in plants. Include information on:

* Why plants need transport systems.
* Structures involved in water and mineral movement from root hairs to xylem, mesophyll cells and stomata.
* Transpiration.
* Transport of sugar through the plant.

I can:

* State that sugar is transported around the plant in phloem.
* State that phloem is made of living cells.
* State that sugar is transported up and down the plant in living phloem cells.
* Explain that phloem contains sieve plates (which control the flow of sugar in phloem).
* Explain that phloem contains companion cells (which control the opening and closing of sieve plates).
* Explain that sugar can be transported up and down phloem due to sieve plates and companion cells.
* Describe the flow of water from the root hair cells, up xylem vessels and into the mesophyll cells of the leaf, finally leaving the plant by diffusing through the stomata.

12 Transport systems – Animals

Blood

In **mammals** the blood acts as a **transport** system to transport **oxygen**, **carbon dioxide** and **nutrients**. Blood contains **red blood cells**, **white blood cells** and **plasma** (fig 12.1).

White blood cells are part of the body's immune system. Red blood cells transport oxygen (fig 12.2). Plasma is the pale yellow liquid part of blood. The main component of plasma is water.

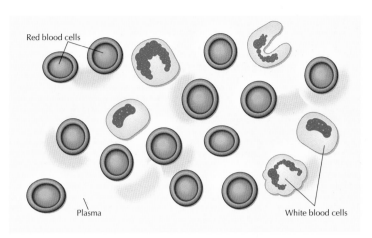

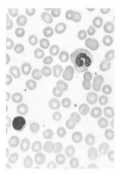

Fig 12.1 *Blood is mostly water, with cells and other substances suspended in plasma*

Fig 12.2 *This picture shows a drop of blood under a microscope. Can you see the two white blood cells in amongst the mass of red blood cells?*

Red blood cells and plasma have important roles in transporting substances around the body. Red blood cells are involved in the transport of oxygen and carbon dioxide.

Substances such as glucose and carbon dioxide dissolve in plasma and are transported around the body in this way (fig 12.3).

The average human adult has 5 litres of blood circulating around their body in blood vessels. The human body takes in useful substances which need to be transported to other parts of the body. It also produces waste, which must be removed. Blood allows substances to be transported from one place to another.

The table shows some substances that are carried in the blood.

Hint

Red blood cells are much smaller than white blood cells. They are also present in the blood in larger numbers.

Substance	Role within the body	Carried from	Carried to
Oxygen	Required to release energy from food	Lungs	All cells of the body
Carbon dioxide	Waste product of respiration	All cells of the body	Lungs
Nutrients (e.g. glucose and amino acids)	Required energy release and growth	Small intestine	All cells of the body

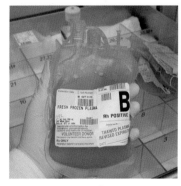

Fig 12.3 *Occasionally, patients require plasma transfusions. This picture shows plasma that has been separated from blood cells received from a blood donor*

Make the link – Biology

Hormones and antibodies travel from one place to another in the bloodstream (see pages **117** and **179**).

Make the link – Health and Wellbeing

Blood transfusions may be given for a number of reasons, for example accidents or disease. Blood transfusions require donations from the public.

Biology in context

If a blood vessel in your body becomes damaged, blood can escape the circulatory system. We can see the effects of

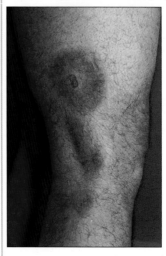

this and know it as bruising. In order for blood to transport substances around your body, it must stay within blood vessels. If a blood vessel becomes damaged, red blood cells start to leak out.

The bruising we see is caused by the red blood cells that have escaped the blood vessel. These red blood cells are removed by white blood cells. During this process they change colour, causing the characteristic colour changes seen as a bruise heals (fig 12.4).

Fig 12.4 *The colour of a bruise is caused by red blood cells that have escaped the circulatory system*

GO! Activities

Activity 2.12.1 Working individually

1. The blood of mammals contains three different parts. List these parts.

2. Name the two parts of blood that are involved in the transport of substances around the body.

3. List three substances that are transported around the body in the blood.

Activity 2.12.2 Working individually

Copy the blood drop picture onto a piece of A5 paper. Fill it with information and diagrams about the role of blood and substances that are transported in it (fig 12.5).

Activity 2.12.3 Working in pairs

Search the web to find information and videos about how bruises are made and how the human body heals itself.

Activity 2.12.4 Working individually

Read the following passage which has been adapted from https://www.scotblood.co.uk/about-blood/journey-of-donated-blood/

Fig 12.5 *Blood drop*

Answer the questions which follow.

Blood donated in Scotland is taken to laboratories, where it is tested and separated into red blood cells, plasma and platelets. The blood is also screened for pathogens.

Donated blood is sent to one of the 39 blood banks in Scotland. Fresh blood is refrigerated to allow it to be used for up to 35 days after donation. Platelets must be given to the recipient within 7 days. Frozen plasma can last for 3 years.

Medical staff, including doctors, nurses, midwives and laboratory staff, receive training in safe transfusion practice.

1. Name three components of blood.
2. Suggest why blood is refrigerated.
3. State the number of times longer fresh blood lasts compared to platelets.
4. Suggest a reason why frozen plasma lasts longer than fresh blood.
5. Explain the meaning of the term "safe transfusion practice".

I can:

- State that the blood of mammals contains plasma, red blood cells and white blood cells.   ◯ ◯ ◯
- State that plasma transports nutrients such as glucose around the body.   ◯ ◯ ◯
- State that red blood cells transport oxygen and carbon dioxide around the body.   ◯ ◯ ◯
- State that white blood cells are part of the immune system.   ◯ ◯ ◯

Red blood cells

Red blood cells are involved in the transport of **oxygen** around the body. They are **specialised** cells. Being specialised allows them to carry out their function **efficiently**. Red blood cells are very numerous in the blood of mammals (fig 12.8). Three ways that they are specialised for their function include:

1. They have a **biconcave shape**.

Their special biconcave shape allows them to be flexible and squeeze through narrow blood capillaries. It also allows them to have a large surface area. Having a large surface area allows diffusion to take place quickly (fig 12.6).

Make the link – Biology

Specialisation is the process whereby a cell becomes suited to its specific function (see page **100**).

Haemoglobin is a protein (see page **48**).

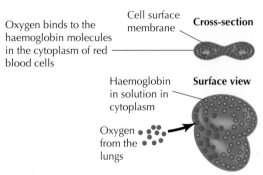

Oxygen binds to the haemoglobin molecules in the cytoplasm of red blood cells

Cell surface membrane

Cross-section

Haemoglobin in solution in cytoplasm

Surface view

Oxygen from the lungs

Fig 12.6 *The biconcave shape of the red blood cell gives it a larger surface area and the presence of haemoglobin allows the efficient transport of oxygen*

2. They contain **no nucleus**.

Red blood cells do not contain a nucleus. This allows them to contain the maximum mass of haemoglobin (fig 12.7).

3. They contain **haemoglobin**.

Haemoglobin is a protein that helps red blood cells to transport **oxygen** to body cells **efficiently** in the form of **oxyhaemoglobin**.

Haemoglobin picks up oxygen in the lungs. When oxygen becomes bound to haemoglobin, a complex called oxyhaemoglobin is formed (fig 12.6). Oxygen is released from the oxyhaemoglobin in the tissues to be used for respiration. An equation can be used to show this reaction:

oxygen + haemoglobin ⇌ oxyhaemoglobin

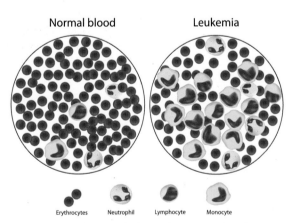

Normal blood   Leukemia

Erythrocytes   Neutrophil   Lymphocyte   Monocyte

Fig 12.7 *Red blood cells do not have a nucleus but white blood cells do*

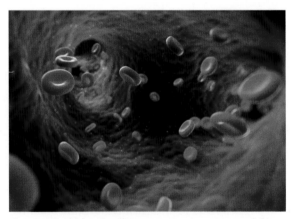

Fig 12.8 *Red blood cells transport oxygen around the body. Haemoglobin makes the transport of oxygen more efficient*

🟢 Activities

Activity 2.12.5 Working individually

1. State the function of red blood cells in the body.
2. List three specialisations of red blood cells which allow them to carry out their function efficiently.
3. Explain how each of these specialisations allows red blood cells to carry out their function efficiently.
4. Name the protein that red blood cells contain that allows them to carry out their function.
5. State the name of this protein when it has oxygen attached to it.
6. Copy and complete the following word equation:

   o_____ + haemoglobin ⇌ o_____

Activity 2.12.6 Working individually

If a person moves from low altitude to high altitude, the number of red blood cells in their body will increase. At high altitude there is a lower concentration of oxygen in the air, so the body adapts to ensure the cells will receive an adequate oxygen supply. The table below shows the change in number of blood cells in a group of mice after they were moved from 1000 m above sea level to 4000 m.

Time at 4000 m (days)	Average number of red blood cells per mm³ blood (millions)
0	7.5
5	9.5
10	10.5
15	11.0
20	11.0

1. On a piece of A5 graph paper, draw a line graph to show the results shown in the table above.
2. State what has happened to the number of red blood cells in the mice transferred to a high altitude.
3. Explain how this change could be advantageous to the mice.
4. If a person is moved from a low altitude to a high altitude quickly their breathing rate may change. Explain why this might happen.
5. Some athletes choose to train at high altitude. Explain how this might give them an advantage during competition.

🌴 Biology in context

Sickle-cell anaemia is a blood disorder that causes red blood cells to have an abnormal shape. The sickle cells can clog blood vessels, leading to pain. The sickle cells also have a shorter lifespan than healthy red blood cells, which can lead to anaemia. Symptoms of anaemia include breathlessness and tiredness, because oxygen supply to the body cells is reduced. There is no cure for sickle-cell anaemia but symptoms can be managed with medications (fig 12.9).

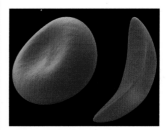

Fig 12.9 *Sickle cells have a very different shape from healthy red blood cells*

🔍 Hint

Remember, haemoglobin becomes oxyhaemoglobin in the lungs. Oxyhaemoglobin becomes haemoglobin in body cells when oxygen passes to body cells.

(Continued)

Activity 2.12.7 Working in pairs

A student used a microscope to look at some blood cells.

Figure 12.10 was identified as a red blood cell.

Explain whether you agree or disagree with the student.

Make a drawing of a red blood cell and label the specialisations which allow it to carry oxygen efficiently.

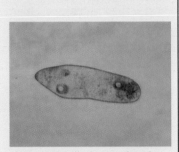

Fig 12.10 *Cell observed and identified as a red blood cell*

I can:

- State that red blood cells are specialised to their function.

- State that the function of red blood cells is to carry oxygen efficiently around the body of mammals.

- State that red blood cells are specialised as they have a biconcave shape, they contain no nucleus but they do contain haemoglobin.

- Explain that having a biconcave shape increases the surface area for carrying oxygen.

- Explain that containing no nucleus allows the red blood cells to contain the maximum mass of haemoglobin.

- Explain that haemoglobin is a protein which joins to oxygen so that oxygen can be transported around the body in the form of oxyhaemoglobin.

- State the following equation: oxygen + haemoglobin $\rightleftharpoons$ oxyhaemoglobin

White blood cells

Fig 12.11 *Microscopic, pathogenic bacteria and viruses*

White blood cells are a part of the **immune system**. The immune system is needed to **destroy** invading **pathogens**. Pathogens are **microorganisms** which cause disease. Some **bacteria**, **fungi** and **viruses** can be classified as pathogens (fig 12.11).

People infected by pathogens may feel very unwell. There are **two main types** of white blood cells which help to defend against pathogens in two different ways. The first type of white blood cell is called a **phagocyte** and the second type is called a **lymphocyte**.

Phagocytes

Phagocytes circulate around the body. They detect chemicals released from pathogens which attract phagocytes to the pathogen. Phagocytes can pass through the walls of blood vessels and into the surrounding tissues (fig 12.12).

Phagocytes will **engulf** the pathogen (fig 12.13). They will then **digest** the pathogen. This process of engulfing followed by digesting is known as **phagocytosis** (fig 12.14).

1 – Phagocyte detects chemicals released by pathogen
2 – Phagocyte squeezes out of blood vessel wall
3 – Phagocyte moves towards the pathogen
4 – Phagocyte engulfs the pathogen

Fig 12.12 *The immune systems responds to a sharp object breaking the surface of the skin allowing pathogens to enter the body*

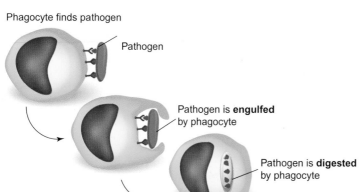

Phagocyte finds pathogen

Pathogen

Pathogen is **engulfed** by phagocyte

Pathogen is **digested** by phagocyte

Fig 12.14 *The process of phagocytosis*

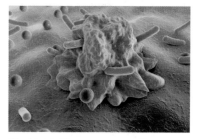

Fig 12.13 *Phagocyte engulfing pathogens*

Lymphocytes

Lymphocytes are a type of white blood cell which produce chemicals to **destroy** pathogens. The chemicals that they produce are called **antibodies** (fig 12.15). Antibodies destroy pathogens by binding to the pathogens and causing them to lump together in a clump. This clump can be more easily engulfed by phagocytes.

Each type of lymphocyte produces only one type of antibody (fig 12.16). Each antibody can destroy only one type of pathogen. This means that every antibody is **specific** to one **particular** pathogen. Antibodies are specific due to their shape at the site where they bind onto pathogens. This shape means that they can only bind onto one pathogen (fig 12.17).

Fig 12.15 *Antibodies are proteins produced by lymphocytes used to destroy pathogens*

Fig 12.16 *Lymphocyte releasing one specific type of antibody*

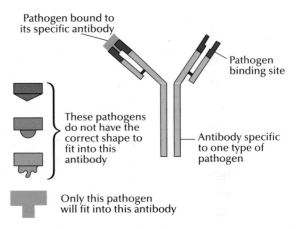

Fig 12.17 *The part of the antibody which binds onto the pathogen has a shape that is specific to only one pathogen*

Labels on figure: Pathogen bound to its specific antibody; Pathogen binding site; These pathogens do not have the correct shape to fit into this antibody; Antibody specific to one type of pathogen; Only this pathogen will fit into this antibody

☿ Biology in context

Blood tests are one of the most commonly used types of medical test. They can be used to diagnose a range of medical conditions. A blood test takes a doctor, nurse or a phlebotomist (a person who specialises in taking blood samples) only a few minutes to carry out (fig 12.18). Blood tests can be used to determine a patient's state of health. A high white blood cell count can indicate that the patient may have an infection. The increase in the number of white blood cells means that the immune system is working hard to destroy the pathogen causing the infection. A high white blood cell count can also indicate that the patient is suffering stress.

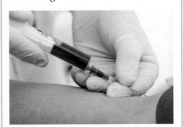

Fig 12.18 *Blood samples are taken to help doctors diagnose their patient's symptoms*

⚛ Make the link – Biology

You learned that enzymes are specific when you studied enzymes in Area of study 1. Antibodies are also specific. Specific just means that antibodies can only bind to and destroy one particular pathogen.

GO! Activities

Activity 2.12.8 Working individually

1. Give the biological term used to describe disease-causing microorganisms.

2. List three different types of disease-causing microorganisms.

3. Name the type of immune system cells which are responsible for destroying disease-causing microorganisms. State the number of different types of these cells and name them.

4. Name the two processes that occur during phagocytosis.

5. Describe the process of phagocytosis.

6. State the function of lymphocytes.

7. Explain the function of antibodies.

8. Antibodies are said to be specific. Explain the meaning of this statement.

Activity 2.12.9 Working individually

Read the following passage and then answer the questions that follow.

It is possible to help the immune system to produce specific antibodies before a person has been exposed to a pathogen. This is what happens when a person receives a vaccination. Since antibodies are specific, more than one vaccination is needed to provide immunity against a range of different pathogens. People in Scotland can receive vaccines against the pathogens that cause measles, mumps, rubella, meningitis, whooping cough, flu and cervical cancer to name but a few.

A vaccine contains a tiny amount of the inactive pathogen that the person is being vaccinated against. Being inactive means that the pathogen in the vaccine cannot cause the recipient to become ill but it does give the lymphocytes a chance to make antibodies to protect against the disease should the person be exposed to it in the future.

A vaccine can contain dead pathogens, toxins which are produced by the pathogen, tiny pieces of the pathogen or live pathogens which have been treated to make them harmless. When the vaccine is injected into the body it stimulates lymphocytes to make antibodies to protect against the pathogen.

1. Name four diseases that people in Scotland can receive vaccines to protect them against.

2. Explain why people are vaccinated.

3. Describe the possible contents of a vaccine.

4. Explain why it is not possible to receive only one vaccine.

5. Explain how a vaccination protects the recipient against a particular pathogen.

Activity 2.12.10 Working in groups

Go to the following webpage:
http://www.immunisationscotland.org.uk/

Click on the vaccines tab. Each member of the group should research a different disease and should complete a 150-word report. This report should be shared with other members of the group.

You might wish to include information on the following areas:

- What the disease is and how it is caused.
- The main signs or symptoms that the person has been infected by the pathogen.
- The name of the vaccine that is used to give immunity to the pathogen.
- The age that the vaccine is first given and whether it requires a booster vaccination.

❋ Make the link – Health and Wellbeing

You may have learned about the importance of having a balanced diet in health and food technology. A balanced diet ensures that you receive all the nutrients you need to keep you as healthy as possible.

As part of a balanced diet you should aim for five-a-day. This is because fruit and vegetables contain vitamins. Vitamin A, vitamin B and vitamin D are particularly important in keeping your immune system healthy.

🔍 Hint

Try to remember that antibodies are produced by lymphocytes (auntie lym) and phagocytes carry out phagocytosis.

I can:

- State that white blood cells are part of the immune system and are involved in destroying pathogens.
- State that pathogens are microorganisms which cause disease.
- Give bacteria, viruses and fungi as examples of microorganisms that could act as pathogens.

- State that two types of white blood cells are phagocytes and lymphocytes.
- State that phagocytes carry out the process of phagocytosis.
- Explain that phagocytosis involves the processes of engulfing the pathogen and then digesting it.
- Explain that each lymphocyte produces a specific antibody.
- Explain that each type of antibody is specific to one particular pathogen.
- Explain that antibodies destroy pathogens.

Function of the heart

The heart is a muscular organ (fig 12.19). The heart is located in the chest cavity. It lies between the lungs (fig 12.20). The function of the heart is to pump blood around the body. The heart pumps blood to the lungs, where blood picks up oxygen. The oxygen-rich blood travels back to the heart and is then pumped around the rest of the body to deliver oxygen to body cells. The left side of the heart contains **oxygenated** blood (oxygen-rich) and the right side contains **deoxygenated** blood (lacking oxygen).

Structure of the heart

Heart chambers

The heart has four chambers. The two upper chambers of the heart are called **atria** (singular: **atrium**) and the lower two chambers are called **ventricles**. The atria are receiving chambers. The left atrium receives blood from the lungs. The right atrium receives blood from the body. The ventricles are pumping chambers. The left ventricle pumps oxygenated blood to the body. The right ventricle pumps deoxygenated blood to the lungs (fig 12.21).

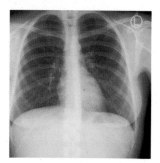

Fig 12.19 *Your heart is about the size of your fist*

Fig 12.20 *The heart is in the centre of the chest*

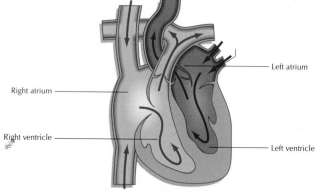

Right atrium

Left atrium

Right ventricle

Left ventricle

Fig 12.21 *This diagram shows a vertical section through the heart. The name and location of the four chambers are identified*

Associated blood vessels of the heart

Heart arteries carry blood away from the heart. Heart veins return blood to the heart. The major blood vessels associated with the heart are:

- **Pulmonary arteries** – arteries that carry deoxygenated blood from the right ventricle to the lungs.

- **Aorta** – an artery that carries oxygenated blood from the left ventricle to the body.

- **Pulmonary veins** – veins that carry oxygenated blood from the lungs to the left atrium.

- **Vena cava** – veins that carry deoxygenated blood from the body to the right atrium (fig 12.22).

- **Coronary arteries** – see below.

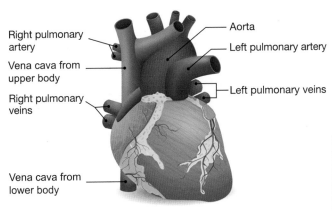

Fig 12.22 *Location of blood vessels associated with the heart*

Coronary arteries

The heart is a muscle, and all muscles require a blood supply. The heart muscle cells require energy to contract and allow the heart to beat.

The heart receives its blood supply from the coronary arteries. The coronary arteries are associated blood vessels. They are actually the first branches of the aorta. The coronary arteries ensure the heart muscle cells have a good supply of glucose and oxygen to carry out respiration and release energy (fig 12.23).

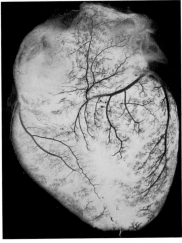

Fig 12.23 *This picture shows a special X-ray of the coronary arteries. The arteries have been filled with dye so that they can be seen*

Heart valves

It is important that blood flows through the heart in one direction only. To ensure this happens, the heart has valves. Valves prevent the backflow of blood (fig 12.24). Valves are found between the atria and ventricles to ensure the blood flows from each atrium to the ventricle below (fig 12.25). Valves are also found at the points where the pulmonary artery and aorta leave the heart.

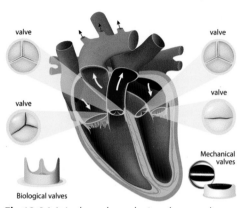

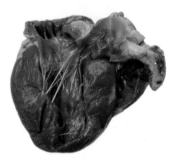

Fig 12.25 *A section through a pig's heart. The string-like structures are tendons, which connect the valves to the heart muscle*

Fig 12.24 *It is the valves closing that you hear if you use a stethoscope to listen to your heart*

Pathway of blood

The circulatory system is made up of the heart and blood vessels. A network of blood vessels extends throughout the body to ensure all cells have an adequate blood supply (fig 12.26). Blood flows through the **heart** in one direction. Blood in the right atrium is forced into the right ventricle. It leaves the heart in the pulmonary arteries and travels to the **lungs**. The blood then returns to the heart in the pulmonary veins. The blood enters the left atrium and is forced into the left ventricle. The blood leaves the heart in the aorta and travels to all the cells of the **body**. Blood returns to the heart in the vena cava where it arrives in the right atrium.

The blood entering the right atrium is **deoxygenated** since it is returning from the body cells and oxygen has been used up. The blood entering the left atrium is **oxygenated** since it is returning from the lungs where it has picked up oxygen (fig 12.27).

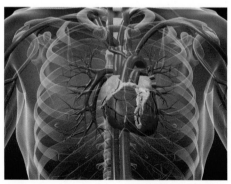

Fig 12.26 *A network of blood vessels extends throughout the body*

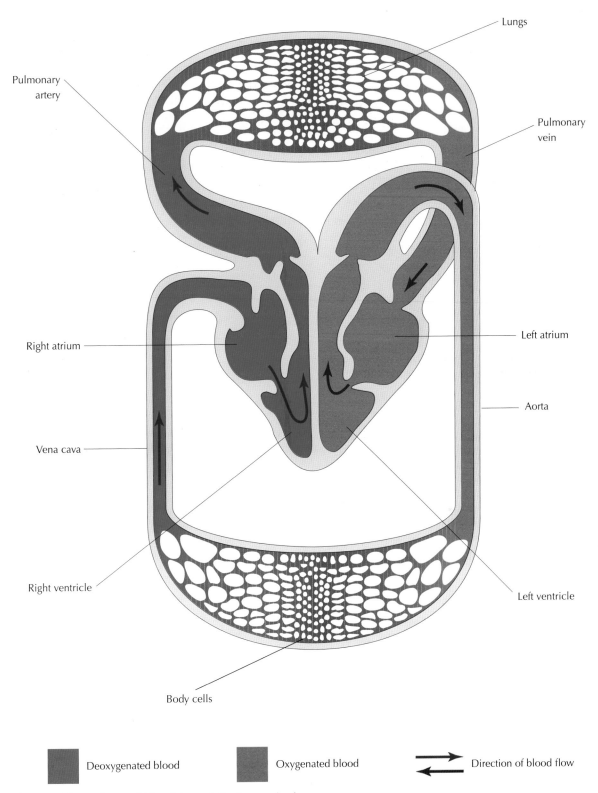

Fig 12.27 *The pathway of blood through the human body*

Make the link – Biology

In Area of study 1, Chapter 6, you learned about respiration. Respiration requires oxygen and glucose. These substances are transported in the blood. The heart also requires oxygen and glucose for respiration since it is made of cells.

Make the link – Health and Wellbeing

You may have learned in food technology the importance of not eating too many fatty foods. This is because too many fatty foods increase the chances of developing coronary heart disease (CHD). The fats are deposited in the walls of arteries and can lead to a condition called atherosclerosis. A person with atherosclerosis is more likely to develop angina. If the artery becomes completely blocked the person might have a heart attack.

Hint

Remember as you look at a diagram of the heart, everything is reversed, so left is right, and right is left! Try looking in the mirror and lifting your right hand – it seems to be on the left of your image!

Biology in context

Heart rate is the number of times the heart beats in 1 minute. The average heart rate at rest is usually between 60 and 80 beats per minute, and this increases dramatically during exercise (fig 12.28). When the heart rate increases, blood is pumped around the body more quickly. This means the hard-working muscles of the body are supplied with more blood containing glucose and oxygen so they can produce more energy by respiration. The increase in heart rate also allows the blood to carry waste products away from the muscles more quickly (fig 12.29).

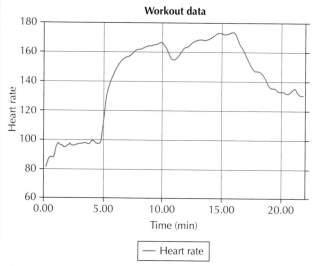

Fig 12.28 *This graph shows how heart rate changes with exercise*

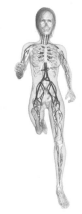

Fig 12.29 *Your cardiovascular system must work much harder when your body is exercising*

GO! Activities

Activity 2.12.11 Working individually

1. Name the upper chambers of the heart.
2. Name the lower chambers of the heart.
3. Copy and complete the sentence below, selecting the correct option from each bracket.

   Oxygenated blood is found in the (left/right) side of the heart, and deoxygenated blood is found in the (left/right) side of the heart.
4. Name the artery that carries blood to the body.
5. Name the veins that return blood from the lungs to the heart.
6. Give the function of the coronary arteries.
7. Copy and complete the flow chart below:

   right a_____ → right v_____ → p_____ a_____ → lungs → p_____ v_____ → left a_____ → left v_____ → a_____ → body → v____ c____ → right a_____.

Activity 2.12.12 Working in pairs

Work in pairs through this activity. Type the following address into your web browser:

www.bhf.org.uk/informationsupport/how-a-healthy-heart-works

Work through the topic to find out more about how the heart works.

Activity 2.12.13 Working in pairs

How to keep your heart healthy.

1. Maintain a healthy diet and weight
2. Do not smoke
3. Get plenty of exercise
4. Avoid stress

Carry out some research and write a couple of sentences to explain how each of the factors above helps to keep your heart healthy.

Activity 2.12.14 Working in pairs

Heart valves may need to be replaced if they become damaged or diseased. Find out about the two types of replacement valves and write a 200-word report giving details of your findings.

Activity 2.12.15 Working in pairs

Work with a partner to carry out some research into CHD.

• Use the internet to find out information about the causes, symptoms, treatment and prevention of CHD.
• Make a leaflet that could be displayed in a doctor's surgery to inform the public about CHD.
• Try to include as much information as possible and make your leaflet look colourful and eye-catching.

(Continued)

Activity 2.12.16 Working in pairs

Imagine you are a red blood cell in the right atrium. Describe the pathway you would take around the body by:

- writing an imaginative story of 200 words

  or
- making a comic strip with six panels.

I can:

- State that the heart contains four chambers.
- Identify the top chambers as atria and the bottom chambers as ventricles.
- State that the left side of the heart is on the right side as it is viewed, and the right side of the heart is on the left side as it is viewed.
- State the heart contains valves to prevent the backflow of blood.
- Identify the location of the heart valves as in between the atria and ventricles and in the aorta and pulmonary artery.
- State that the pulmonary arteries carry blood from the heart to the lungs, and the pulmonary veins carry blood from the lungs back to the heart.
- State that the vena cava carries blood from the body to the heart, and the aorta carries blood from the heart to the body.
- State that the coronary arteries supply the heart muscle with blood containing glucose and oxygen.
- Describe the flow of blood around the body as:

  right atrium → right ventricle → pulmonary arteries → lungs → pulmonary veins → left atrium → left ventricle → aorta → body → vena cava → right atrium.
- State that the left side of the heart contains oxygenated blood and that the right side of the heart contains deoxygenated blood.

Blood vessels

Every person contains a network of tubes inside their bodies that are needed to carry blood around the body. These tubes are called **blood vessels**. The blood vessels and the heart make up the circulatory system (fig 12.30).

There are three different types of blood vessel. These are **arteries**, **veins** and **capillaries**. The structure of each type of blood vessel is suited to its specific function (fig 12.31).

Arteries

Arteries carry blood **away from the heart**. They carry blood which is **under high pressure**. Arteries have **thick, muscular walls** to withstand the high pressure. When the heart pumps blood into an artery, the pressure causes the wall of the artery to push out. When the heart relaxes, the wall rebounds, and this can be felt as your pulse.

Compared to veins, arteries have a **narrow central channel** to carry blood (fig 12.32).

Veins

Veins return blood **back towards the heart**. They carry blood which is under **low pressure**. Veins have relatively **thinner** muscular walls as the blood they transport is under low pressure. The blood in veins is at much lower pressure than the blood in arteries. Compared to arteries, veins have a **wider central channel** to carry blood (fig 12.32).

Veins contain structures called **valves**. Blood can flow through valves in one direction only. Valves **prevent the backflow of blood** (fig 12.33).

Capillaries

Capillaries connect arteries to veins. They branch from arteries and form **networks at tissues and organs**. They reunite into veins, which leave the tissues and organs (fig 12.34).

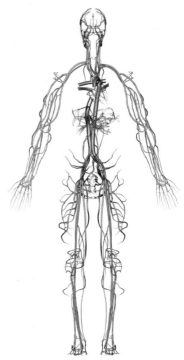

Fig 12.30 *The circulatory system extends throughout the body, ensuring all cells have a good blood supply*

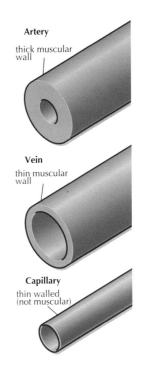

Fig 12.31 *Arteries, veins and capillaries have very different structures which are suited to their function*

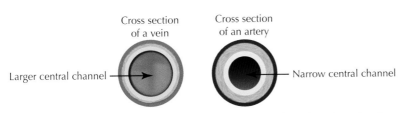

Cross section of a vein

Cross section of an artery

Larger central channel

Narrow central channel

Fig 12.32 *The central channel of an artery is narrower than that of a vein of the same diameter*

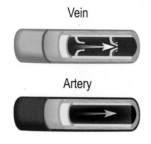

Fig 12.33 *The only type of blood vessel to contain valves is veins. Valves prevent the backflow of blood*

Fig 12.34 *Capillaries are so narrow that red blood cells can only just pass through one by one*

🔍 **Hint**

Remember **a**rteries carry blood **a**way from the heart – **A**rteries **A**way.

🔍 **Hint**

Veins carry blood back to the heart and contain valves.

Capillaries create a **large surface area** for diffusion to occur (fig 12.35). The large surface area allows **materials** to be **efficiently exchanged** between the blood and the tissues. Capillaries have very **thin walls**, which helps the efficient exchange of materials. The walls of capillaries are only one cell thick, which creates a large surface area. This allows diffusion to take place as quickly as possible (fig 12.36).

Fig 12.36 *Capillaries are thin-walled blood vessels. The purple shapes on the diagram are the nuclei of the cells which make the thin walls*

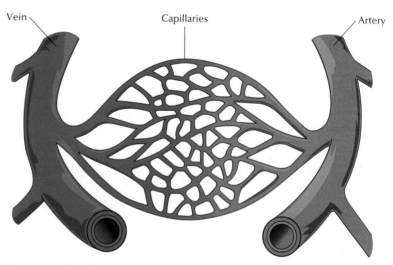

Fig 12.35 *Blood leaves the heart in arteries and then flows into capillaries. Materials are exchanged between the blood in the capillaries and the body cells. Blood then flows back to the heart in veins*

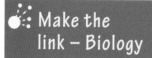
Make the link – Biology

Valves are found in the heart where they prevent the backflow of blood (see page **184**).

Biology in context

Deep-vein thrombosis (DVT) is a blood clot that becomes lodged in a vein. This can happen in any vein in the body, most often in deep leg veins. This can cause pain, swelling and redness of the leg. DVT can affect anyone. However, some people are more at risk than others. For example:

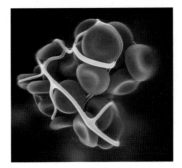

Fig 12.37 *Blood clots are made up of blood cells and other proteins such as fibrin*

- Those who are overweight
- People who have been sedentary for long periods of time, e.g. on a long-haul flight
- People with medical conditions such as heart failure.

If a person develops DVT, they may be given anticoagulation drugs, which prevent clots from forming so easily. They may also be advised to raise their leg when resting to prevent blood pooling in the leg. DVT is more likely to happen if blood is circulating slowly, so it is important to take regular exercise to improve circulation. If a person knows they will be sedentary for a long period of time, they should keep hydrated, perform simple leg exercises and take walks whenever possible to reduce the risk of developing DVT (fig 12.37).

Make the link – Health and Wellbeing

In PE you may have measured your pulse rate before and after exercising. In PE you would do this to work out your recovery time. Recovery time is used as a measure of fitness as it tells you how quickly your heart returns to its resting pulse after exercise. By taking your pulse you are indirectly measuring your heart rate. As blood is pumped through the arteries by the heart you can feel your pulse when the arteries swell.

Activities

Activity 2.12.17 Working individually

1. Name the type of blood vessels which carry blood away from the heart.
2. Describe the walls of arteries compared to the walls of veins.
3. Describe the central channel of arteries compared to the central channel of veins.
4. Compare and contrast the pressure of blood in arteries compared to veins.
5. Name the type of blood vessel which carries blood back to the heart.
6. Name the type of blood vessels which contain valves.
7. Explain the function of valves in blood vessels.
8. Name the type of blood vessels that have walls which are only one cell thick.
9. Explain the importance of capillaries forming networks at tissues and organs.

(Continued)

Activity 2.12.18 Working individually

Copy and complete the table below:

Type of blood vessel	Function	Description of vessel wall	Relative width of central channel	Special features
Artery				
Vein				
Capillary				

Activity 2.12.19 Working in pairs

Carry out some research into DVT. Produce a short slideshow presentation that could be shown before a long-haul flight. Make sure you explain:

- what DVT is
- how it can be prevented
- the treatments that are used if a person develops DVT.

I can:

- State that arteries carry blood away from the heart.
- State that arteries have thick, muscular walls compared to veins.
- State that arteries have a narrow central channel to carry blood when compared to a vein of the same diameter.
- Explain that arteries carry blood under high pressure.
- State that veins carry blood back to the heart.
- State that veins have thinner muscular walls compared to arteries.
- State that veins have a wider channel to carry blood when compared to an artery of the same diameter.
- Explain that veins carry blood under low pressure.
- State that veins contain valves.
- State that valves prevent backflow of blood.
- State that capillaries are thin-walled blood vessels.
- State that capillaries provide a large surface area to allow efficient exchange of materials.
- State that capillaries form networks at tissues and organs to allow efficient exchange of materials.

13 Absorption of materials

Breathing system

When we take a breath in, air moves into our **lungs** through the breathing system. Air containing **oxygen** enters the breathing system through the nose and mouth. Oxygen then passes down a series of breathing tubes till it reaches the lungs (fig 13.1). Oxygen travels within the lungs through a further series of smaller and smaller breathing tubes. At the end of all these breathing tubes there are air sacs called **alveoli** (fig 13.2).

It is in the alveoli that **gas exchange** takes place. Gas exchange allows **oxygen** to enter the blood capillaries. **Carbon dioxide** is exchanged at the same time and so leaves the blood capillaries. When we breathe out, air moves from the alveoli through the breathing tubes until it reaches the nose and mouth and is breathed out. Due to the exchange of oxygen and carbon dioxide in the alveoli the **lungs** are known as the **gas exchange organs**.

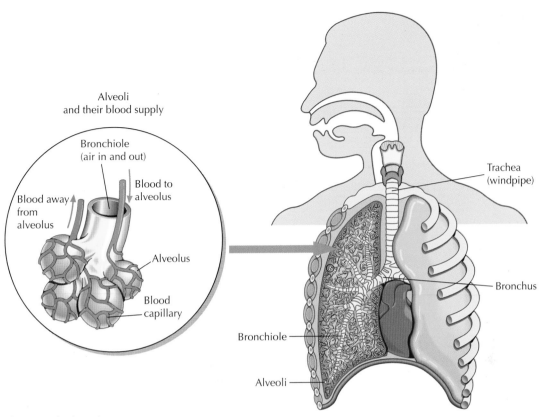

Fig 13.1 *The breathing system*

Alveoli

Alveoli are the site of gas exchange in the lungs. They allow oxygen to diffuse into the blood and they allow carbon dioxide to diffuse out of the blood. All body cells require oxygen for respiration. Carbon dioxide is produced as a waste product of respiration. The lungs contain a **large number of alveoli**. It is important that gas exchange happens quickly and efficiently.

The alveoli allow efficient gas exchange in several ways:

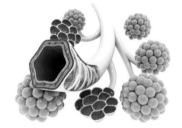

Fig 13.2 *Breathing tubes in the lungs end in tiny air sacs called alveoli*

- There are a **large number of alveoli** in the lungs. This **increases the surface area** available for gas exchange (fig 13.3).

- The alveoli **walls are thin**. This allows rapid **absorption** of oxygen and carbon dioxide (fig 13.4).

- There are **many blood capillaries**. Each alveolus is surrounded by a blood capillary giving it a good blood supply (fig 13.5). Oxygen is absorbed through the thin alveolar walls into blood capillaries. Carbon dioxide is absorbed through the thin alveolar walls from the many blood capillaries into the breathing tubes so it can be breathed out.

> ●ᶜ Make the link –
> Health and
> Wellbeing
>
> Aerobic exercise is recommended to keep your lungs healthy and working efficiently. Aerobic exercise can also be called cardio.

Fig 13.3 *The alveoli create a surface area equal to the size of a tennis court*

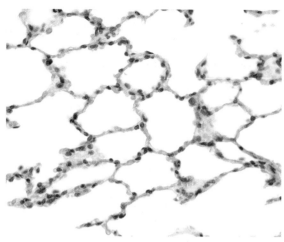

Fig 13.4 *A high magnification section through alveoli showing walls one cell thick*

Fig 13.5 *A medically accurate illustration of alveoli showing many blood capillaries on the surface*

Biology in context

Mechanical ventilation is a treatment used if a person is suffering from respiratory failure. A person with this condition cannot breathe out carbon dioxide leading to it building up in the blood. If the level of carbon dioxide in the blood gets too high the person will become very unwell.

The ventilator blows air into the lungs. This helps breathing as the muscles which make the lungs work are helped to function. The main aim of mechanical ventilation is to reduce the levels of carbon dioxide in the blood. The ventilator also helps to make sure that the patient takes the right number of breaths per minute.

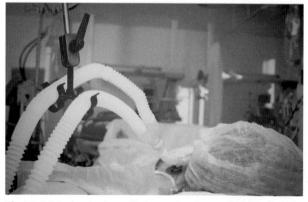

Fig 13.6 *Mechanical ventilation is used to reduce carbon dioxide levels in the blood*

GO! Activities

Activity 2.13.1 Working individually

1. Name the human gas exchange organs.
2. Name the tiny structures found in human gas exchange organs which allow gas exchange to occur.
3. Give the name of both gases exchanged in the lungs and describe the direction they move in.
4. Explain the term 'gas exchange organ'.
5. State three adaptations of lungs that allow them to efficiently exchange gases.

Activity 2.13.2 Working individually

Place two fingers on the front of your neck and move them up and down. The bumps you feel are cartilage which keeps your windpipe open at all times.

Think about the tubes of a vacuum cleaner. List the ways that your windpipe and a vacuum cleaner tube are similar and different.

Activity 2.13.3 Working individually

Search the terms 'lung dissection' and 'lung corrosion model' on the internet to investigate the structure of the lungs.

🔍 Hint

Remember that there are millions of alveoli in the lungs. The lungs are efficient gas exchange structures as there are such a large number of alveoli in each lung.

⚛ Make the link – Biology

Oxygen is required to break down glucose (see page **71**).

Oxygen diffuses from a high concentration in the alveoli to a low concentration in the blood (see page **29**).

🔍 Hint

Remember breathing and respiration are different. Respiration is the process of releasing energy from food. Breathing is the process of bringing air into the lungs.

I can:

• State that lungs are gas exchange organs.

• Explain that oxygen is absorbed into blood capillaries found in the lungs.

• Explain that carbon dioxide is removed from the blood capillaries when they enter the lungs.

• State that the lungs contain a large number of alveoli.

• Explain that the large number of alveoli provide a large surface area for gas exchange.

• Explain that since the alveoli have thin walls this allows rapid gas exchange.

• Describe gas exchange as the process where oxygen is absorbed through alveolar walls into blood capillaries and carbon dioxide is removed from blood capillaries through thin alveolar walls.

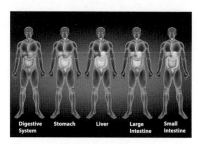

Fig 13.7 *The human digestive system contains many organs which work together to digest food and absorb nutrients*

Digestive system

Digestion is the process which breaks down large, insoluble food molecules into small, soluble food molecules. Soluble food molecules are able to enter the bloodstream to be transported to body cells. Digestion occurs along the digestive system (fig 13.7).

Food enters the digestive system in the mouth and travels to the stomach before entering the small intestine. Digestive enzymes break down food as it travels from the mouth to the small intestine. Digestion is completed in the small intestine. The soluble products of digestion are absorbed into the blood from the small intestine (fig 13.8).

Small intestine

Food spends around 5 hours in the small intestine to allow digestion and absorption to be completed (fig 13.8). Insoluble carbohydrates are broken down into **glucose**. Insoluble proteins are broken down into **amino acids**. Insoluble fats are broken down into **fatty acids** and **glycerol**. These products of digestion are all soluble meaning that that they can be absorbed by the small intestine. The **small intestine allows nutrients from food** to be **absorbed** by **villi**.

To allow efficient absorption of the products of digestion the internal surface of the small intestine is covered with millions of projections called villi (singular: villus). Figure 13.9 shows the internal surface of the small intestine.

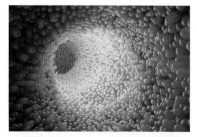

Fig 13.9 *The internal surface of the small intestine is lined with millions of small projections called villi*

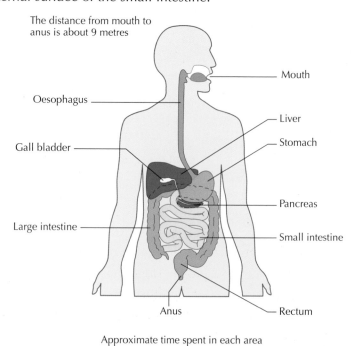

The distance from mouth to anus is about 9 metres

- Oesophagus
- Gall bladder
- Large intestine
- Anus
- Mouth
- Liver
- Stomach
- Pancreas
- Small intestine
- Rectum

Approximate time spent in each area

■ 10 seconds ■ 1–5 hours □ 5 hours ■ 14–24 hours

Fig 13.8 *The time spent by food in each part of the human digestive system*

Villi contain adaptations which make them suitable for their function. Each villus:

- is **thin walled**

The wall of each villus is only one cell thick. This allows digested food to be absorbed very quickly from the small intestine into the villus (fig 13.10).

- is present in **large numbers**

The lining of the small intestine contains a very large number of villi. This allows the villi to have a **large surface area** for absorption of digested food (fig 13.9).

- contains a **network of blood capillaries**

Inside each villus is a blood capillary. This gives the villi a good blood supply and allows quick transport of digested food to body cells. The blood capillary absorbs **glucose** and **amino acids** (fig 13.11).

- contains a **lacteal**

The lacteal absorbs **fatty acids** and **glycerol**. It is part of the lymphatic system which is used to transport these products of fat digestion around the body (fig 13.11). Lacteals increase the surface area for absorption.

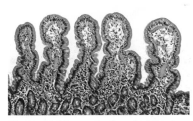

Fig 13.10 *The wall of each villus is thin to allow rapid absorption of digested food*

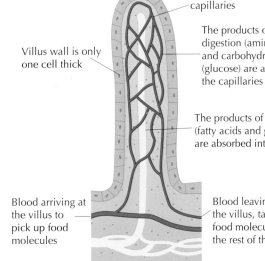

Contains a network of blood capillaries

The products of protein digestion (amino acids) and carbohydrate digestion (glucose) are absorbed into the capillaries

The products of fat digestion (fatty acids and glycerol) are absorbed into the lacteal

Villus wall is only one cell thick

Blood arriving at the villus to pick up food molecules

Blood leaving the villus, taking the food molecules to the rest of the body

Fig 13.11 *A villus*

⊙! Activities

Activity 2.13.4 Working individually

1. State the location where digested food is absorbed in the small intestine.
2. Name four features of the small intestine which allow it to be efficient in its function.
3. Explain how each of these features allows the small intestine to be efficient in its function.
4. Give the products of carbohydrate digestion.
5. Give the products of protein digestion.
6. Give the products of fat digestion.
7. State the part of the small intestine where each of these products are absorbed.

(Continued)

🔍 Hint

Remember that the small intestine contains a very large number of villi. Each of these villi have four adaptations which allow them to absorb nutrients quickly and efficiently. Make sure that you can list and explain these adaptations.

⁙ Make the link – Biology

The wall of each villus is only one cell thick. Nutrients only have to diffuse through one cell membrane to enter the villus. This allows diffusion of nutrients into the villi to be very quick.

Make the link – Health and Wellbeing

In food technology, you may consider the dietary needs of different individuals. A person with coeliac disease must follow a strict gluten-free diet.

Activity 2.13.5 Working individually

- Draw a diagram to show the structure of a villus. Include on your diagram:
  - Labels naming each part of a villus.
  - Information explaining the role each part of the villus plays in the absorption of nutrients.
  - Locations where specific nutrients are absorbed.

Remember to use colour in your diagram.

Activity 2.13.6 Working in pairs

- A person who is unaware that they have coeliac disease can cause damage to the villi in their small intestine.
- One symptom they may show is weight loss. Explain how this symptom relates to the damage caused by the disease.
- Carry out some research on anaemia. Explain how coeliac disease may cause anaemia.

I can:

- State that the inside lining of the small intestine contains a large number of villi.
- State that nutrients from food are absorbed into the villi.
- Explain that the large number of villi give the small intestine a large surface area for absorption.
- State that each villus has a thin wall, a blood capillary and a lacteal.
- Explain that the thin walls allow the small intestine to absorb nutrients very quickly.
- Explain that each villus containing a blood capillary gives a large surface area for absorption of the products of carbohydrate and protein digestion.
- Explain that each villus containing a lacteal gives a large surface area for absorption of the products of fat digestion.
- Explain that amino acids and glucose are absorbed into the blood capillary.
- Explain that fatty acids and glycerol are absorbed into the lacteal.

Biology in context

Coeliac disease is an inherited disorder where sufferers are intolerant to a substance called gluten, which is found in wheat. If a person with coeliac disease eats wheat, this can result in a shortening of their villi. The damage to the villi is caused by the body's own immune system attacking them. This type of condition is known as an autoimmune disease.

Fig 13.12 *People with coeliac disease must avoid wheat products or risk causing damage to their villi*

Symptoms of coeliac disease include diarrhoea, abdominal pain, weight loss and lack of energy. If a person shows a combination of these symptoms, a blood test or biopsy from their small intestine can confirm if they are suffering from coeliac disease. Currently, the only treatment for this condition is to follow a gluten-free diet, which allows the intestines to heal and eases the symptoms associated with the disease (fig 13.12).

Absorption of useful substances

Substances needed by cells for **respiration** are **delivered** to them by the **bloodstream**. **Oxygen** is **absorbed** in the alveoli and diffuses into the blood capillaries which surround the alveoli. The absorption of oxygen happens quickly as the walls of the alveoli are only one cell thick (fig 13.13).

Nutrients from food also have to be delivered to cells for use in respiration. These nutrients are absorbed by the villi in the small intestine. The absorption of nutrients happens quickly as the walls of the villi are only one cell thick. **Glucose** is an example of a nutrient which is used in respiration (fig 13.14).

Fig 13.13 *Oxygen diffuses into the blood capillary to be delivered to respiring cells. Carbon dioxide diffuses out of the blood capillary to be breathed out as it is a waste substance*

Absorption of waste materials

The process of respiration creates **carbon dioxide**. Carbon dioxide is a **waste material**. If it builds up in cells it changes the pH of the cell. This may cause enzyme-controlled reactions to work inefficiently. Body cells must **remove** waste carbon dioxide. They do this by passing carbon dioxide into the **bloodstream**. The bloodstream transports carbon dioxide to the alveoli. The thin walls of the alveoli allow carbon dioxide to be absorbed quickly into the breathing system so that carbon dioxide can be breathed out (fig 13.15).

Make the link – Biology

In Area of study 2, Chapter 12, you learned about the structure of blood vessels. Do you understand that arteries and veins have thick walls? This is why they are unable to absorb useful substances – it would take too long for the substances to pass through their thick walls.

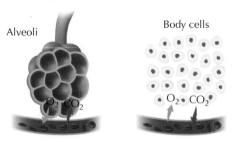

Gas exchange

Fig 13.15 *Body cells produce carbon dioxide when they respire. Carbon dioxide is removed from the blood in the alveoli*

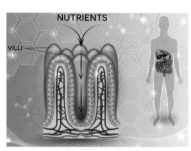

Fig 13.14 *Glucose is one of the nutrients absorbed by the villi which is needed for respiration*

🔍 Hint

Remember that the walls of the alveoli are one only one cell thick. This allows efficient absorption of oxygen.

⚛ Make the link – Biology

In Area of study 1, Chapter 6, you learned about respiration. Read over this topic to refresh your memory of the substances needed by body cells for respiration and why body cells need to carry out respiration.

⚛ Make the link – Health and Wellbeing

In PE you might have completed the bleep test. When you have to stop the bleep test you should have noticed that your breathing rate was higher than when you are resting. One of the reasons for this is because your circulatory system is working harder to remove carbon dioxide from your blood. What would the other reason be?

🟢 Activities

Activity 2.13.7 Working individually

1. State two substances required by body cells for respiration.
2. Explain how these substances are delivered to cells.
3. Give a waste substance produced by respiration.
4. Explain how this waste substance is removed from body cells.

Activity 2.13.8 Working in pairs

The normal pH of blood is between pH 7.35 and pH 7.45. If the blood pH becomes lower than pH 7.35, the person could be suffering from a condition called respiratory acidosis.

Respiratory acidosis occurs when the lungs cannot remove carbon dioxide (CO_2) efficiently. The increase in carbon dioxide concentration in the blood makes the blood too acidic.

Use the internet to research respiratory acidosis and write a report. You should include:

- The causes of respiratory acidosis
- The forms of respiratory acidosis
- The symptoms of respiratory acidosis
- The treatment for respiratory acidosis.

Your report should be around 250 words.

Activity 2.13.9 Working in groups

Carry out an investigation into the effect of exercise on breathing rate. The aim of the investigation is to find out if exercise increases breathing rate.

Discuss the aim with members of your group and then design your experiment.

You should consider:

- Are there any health and safety issues?
- The method you should use to ensure your experiment is valid:
    1. How long should the exercise last?
    2. What type of exercise should you carry out?
    3. Who will be the subject of the experiment?
    4. How will you be able to compare the results?
    5. How will you obtain and record your results?

It would be anticipated that you will find out that exercise increases breathing rate. Use your biological knowledge to explain why exercise increases breathing rate.

I can:

- State that oxygen and nutrients must be absorbed into the bloodstream.
- State that the bloodstream delivers oxygen and nutrients to respiring cells.
- State that waste materials must be removed from the bloodstream.
- State that carbon dioxide is a waste material.

🌳 Biology in context

A pulse oximetry test measures how much oxygen your blood is carrying. It is a simple test that involves attaching a sensor to the skin. The test measures the oxygen saturation of the blood (fig 13.16).

Fig 13.16 *A pulse oximetry test is used to determine the blood oxygen level*

People with lung conditions may have to monitor their blood oxygen level. If it is too low the cells in the body will not be able to respire efficiently. A very low blood oxygen level could make the person unable to think properly and they may appear confused.

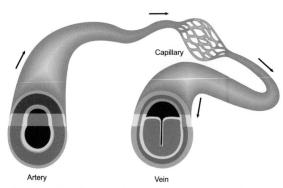

Capillary

Artery   Vein

Fig 13.17 *Capillaries link arteries to veins. Their function is to allow exchange of materials*

Capillary networks

Capillaries are the smallest type of blood vessel. They have very thin walls. Their walls are only one cell thick. Capillaries link arteries to veins (fig 13.17). The function of capillaries is to allow nutrients and oxygen to diffuse into body cells quickly and to allow wastes, such as carbon dioxide, to diffuse out of body cells quickly.

A group of specialised cells working together to carry out the same function is called a **tissue**. Tissues contain **capillary networks**. A capillary network is the term used to describe the many branches capillaries form when they enter the tissue (fig 13.18).

Capillary networks inside tissues allow the **exchange** of materials at a **cellular level**. This is because each cell lies very close to a capillary. The distance that materials have to diffuse over is very small. This allows diffusion to occur rapidly.

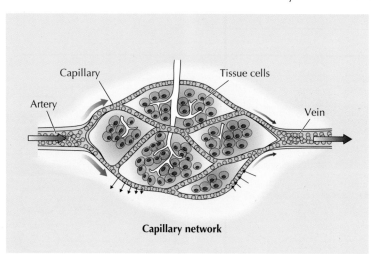

Capillary   Tissue cells

Artery   Vein

Capillary network

Fig 13.18 *Capillary networks are found inside tissues. They allow rapid diffusion of substances since they lie very close to tissue cells*

GO! Activities

Activity 2.13.10 Working individually

1. State what a tissue is.

2. Describe a capillary.

3. Explain the difference between a capillary and a capillary network.

4. Explain how capillary networks allow the exchange of materials at a cellular level.

Activity 2.13.11 Working individually

Draw and label a capillary network. Include the following labels:

- Artery transporting blood into capillary network
- Capillary network
- Vein transporting blood out of the capillary network
- Tissue cells
- Oxygen diffusing into tissue cells
- Carbon dioxide diffusing out of tissue cells
- Nutrients diffusing into tissue cells

Activity 2.13.12 Working in pairs

The following diagram shows some blood vessels on a person's face (fig 13.19).

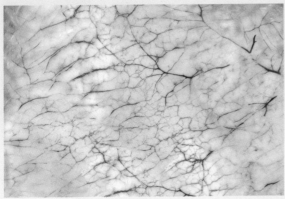

Fig 13.19 *Visible blood vessels*

Use your biological knowledge to explain how you know this is a capillary network.

I can:

- State that a tissue is a group of specialised cells working together to carry out the same function.

- State that capillary networks form in tissues.

- Explain that capillary networks allow rapid exchange of materials at a cellular level due to diffusion.

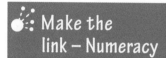

Make the link – Numeracy

In maths you may have learned how to use a ruler to measure distances in centimetres or millimetres. You would not be able to measure the diameter of a capillary with a ruler. The diameter of capillaries ranges in length from just 5 to 10 micrometres.

Biology in context

The Forth Bridges were built to allow people living outside the city centre a more efficient commute to Edinburgh. The bridges provide the quickest link between Fife and the city of Edinburgh. Without the bridges, people would have to drive a much longer distance. The long commute might make working in Edinburgh too time consuming for them. Capillaries do a similar job in the body. They link arteries to veins and ensure that all body cells lie as close as possible to the transport network. This ensures that cells receive oxygen and glucose as quickly as possible. It also ensures that they get rid of waste efficiently (fig 13.20).

Fig 13.20 *The Forth Bridges provide a link to make commuting into Edinburgh more efficient*

Absorbing surfaces

Absorbing surfaces are surfaces which allow the exchange of **materials** between body cells, or between body cells and a transport system. Surfaces which are involved in absorbing materials have certain **features in common**. These features increase the **efficiency of absorption**. These include:

- **a large surface area**
- **thin walls**
- **an extensive blood supply**

Feature which increases efficiency of the absorbing surface	Reason that the feature increases efficiency of absorption	Absorbing surface:
Large surface area	Having a large surface area allows more of the material which has to be absorbed to be in contact with the absorbing surface.	• Villi – having a large number of villi in the small intestine increases the ability of the small intestine to absorb digested food. • Alveoli – having a large number of alveoli in each lung increases gas exchange in the lungs.
Thin walls	Thin walls reduce the distance that a material must diffuse. This makes the process of diffusion faster.	• Villi – each villus is surrounded by a wall which is only one cell thick. • Alveoli – each alveolus is surrounded by a wall which is only one cell thick.
Extensive blood supply	The extensive blood supply means that materials which are absorbed can enter the bloodstream more quickly. The materials are transported around the body in the bloodstream.	• Villi – each villus contains a blood capillary inside. Since there are a large number of villi this results in an extensive blood supply. • Alveoli – each alveolus contains a blood capillary on its surface. Since there are a large number of alveoli this results in an extensive blood supply.

 Biology in context

Malnutrition is a term used to describe the lack of proper nutrition some people suffer. Malnourished people tend to lose weight without trying. It can be caused when a person does not have enough food to eat – the absorbing surfaces of the small intestine do not have enough food to absorb. This can be seen in areas where famine is common. It can also be caused by not being able to make use of the food which has been eaten. This could be caused by absorbing surfaces of the small intestine failing to work properly. Malnutrition can also occur in people who are anorexic. This is because people who are anorexic do not eat enough of the types of food needed to stay healthy. The absorbing surfaces of the small intestine work properly but there isn't enough food to absorb.

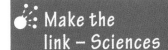 Make the link – Sciences

You may have studied radiation of heat in physics and know that dark and matte surfaces are better at absorbing heat than light and shiny surfaces. Absorbing means the same in both biology and physics. Absorbing means to take something in.

🔍 Hint

If a surface carries out absorption it will possess three features. These are a large surface area, an extensive blood supply and thin walls.

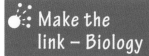 Make the link – Biology

Absorption of materials in the body happens as a result of diffusion. Read over Area of study 1, Chapter 2, to revise diffusion and how this process occurs in the body.

GO! Activities

Activity 2.13.13 Working individually

1. State three features that each absorbing surface possesses.
2. Describe how each of these features aids the absorbing surface to carry out its function.
3. Explain why absorbing surfaces must contain these features.

Activity 2.13.14 Working in pairs

Produce a leaflet to compare and contrast absorbing surfaces in the lungs and small intestine. Your leaflet should include:

1. A labelled diagram of each absorbing surface.
2. The location in the body of each absorbing surface.
3. The reason why the body has these absorbing surfaces.
4. Any differences in the location of the blood supply.
5. Any structures found in an absorbing surface which is not present in the other absorbing surface.

Activity 2.13.15 Working in pairs

There are other absorbing surfaces in the body. For example, those found in the large intestine and in the kidneys. You may want to research the ways which these absorbing surfaces are like those found in the lungs and small intestine. Think about the features that these surfaces must have to allow them to carry out their function efficiently – you should know some already.

I can:

- State that absorbing surfaces have features in common.
- List having a large surface area, an extensive blood supply and thin walls as features they have in common.
- State that these features allow them to increase the efficiency of absorption.

Multicellular organisms – Review questions

Section A

1. Which of the following shows terms listed in order of their hierarchy from lowest to highest in a multicellular organism?

    A   systems → organs → tissues → cells

    B   organs → cells → tissues → systems

    C   tissues → cells → organs → systems

    D   cells → tissues → organs → systems

2. Typical timings of the stages of mitosis are shown in the table below.

Stage	1	2	3	4
Time (minutes)	87	34	26	53

    What percentage of the total time for mitosis is taken by stage 3?

    A   13

    B   26

    C   56

    D   74

3. Which part of the CNS controls heart and breathing rate?

    A   Spinal cord

    B   Cerebellum

    C   Medulla

    D   Cerebrum

4. The diagram below shows the structure of a flower.

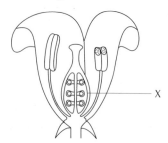

Which line in the table below correctly identifies X and the type of gamete it produces?

	Name of X	Type of gamete produced
A	Anther	Male
B	Anther	Female
C	Ovary	Male
D	Ovary	Female

5. Sperm production in humans is controlled by two hormones, S and T.

As levels of S rise, sperm production increases.

As levels of T rise, sperm production decreases.

Which of the graphs below shows the change in hormone levels of a man whose sperm production is increasing?

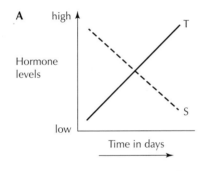

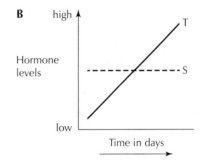

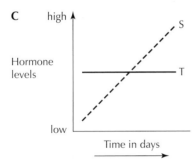

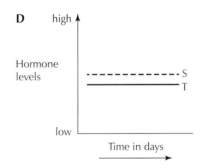

6. In humans, the allele for free earlobes is dominant to the allele for attached earlobes. Two parents both have free earlobes. Their child has attached earlobes.

What is the best explanation for this pattern of inheritance?

A   The parents are heterozygous for the earlobe alleles.

B   The parents are homozygous for the earlobe alleles.

C   One parent is homozygous, and the other is heterozygous, for the earlobe alleles.

D   The child has inherited the attached earlobes directly from a grandparent.

7. The diagram below shows a cross section through a leaf.
    Which letter correctly identifies a spongy mesophyll cell?

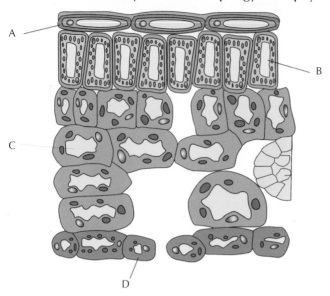

8. The diagram below shows a type of tissue found in a plant.

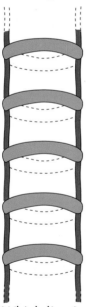

Which line in the table correctly identifies this tissue and its function?

	Tissue name	Function
A	Xylem	Transports water
B	Phloem	Transports sugar
C	Xylem	Transports sugar
D	Phloem	Transports water

9. The diagram below shows a villus. Which letter identifies the blood capillary?

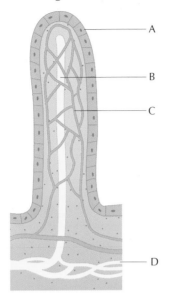

10. A group of students investigated the effect of increasing carbon dioxide concentration in the air on breathing. The volume of air inhaled by one student was measured as the carbon dioxide concentration in the air increased. The results are shown in the graph below.

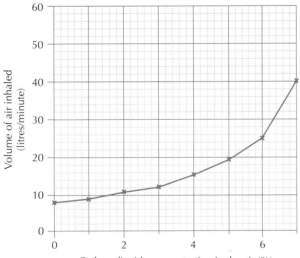

When the carbon dioxide concentration in the air is increased from 1% to 6%, the volume of air inhaled increases by

A   9 litres/minute

B   15.5 litres/minute

C   20 litres/minute

D   25 litres/minute

Section B

1. The diagram below contains some of the stages of mitosis.
   Describe **Stages 2** and **5** in the spaces provided.

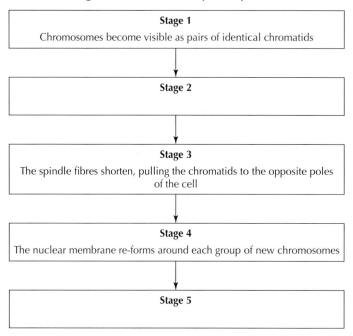

| **Stage 1** |
| Chromosomes become visible as pairs of identical chromatids |

| **Stage 2** |

1

| **Stage 3** |
| The spindle fibres shorten, pulling the chromatids to the opposite poles of the cell |

| **Stage 4** |
| The nuclear membrane re-forms around each group of new chromosomes |

| **Stage 5** |

1

2. **(a)** Give the function of a reflex action.

_____

1

   (b) The diagram below shows the flow of information from a receptor to an effector in a reflex action.

   Complete the diagram by inserting the names of the missing neurons in the correct order.

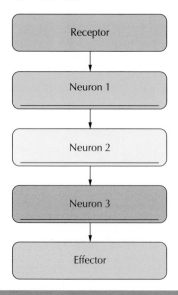

Receptor

Neuron 1
_____

Neuron 2
_____

Neuron 3
_____

Effector

2

3. **(a)** Name the type of glands that produce hormones.

    _____ 1

    (b) Describe how hormones travel from the glands where they are produced to the place where they have an effect.

    _____ 1

4. **(a)** The diagram below shows the control mechanisms used by the body to regulate blood glucose levels.

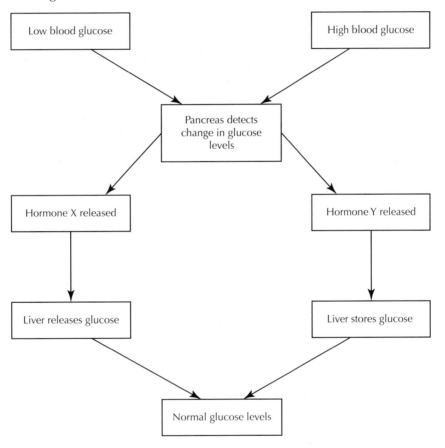

    Name hormones X and Y.

    Hormone X _____ 1

    Hormone Y _____ 1

(b) People who suffer from diabetes cannot regulate their blood glucose levels. A potential treatment for type 1 diabetes involves the use of stem cells.

Define the term stem cells.

_____

_____ **1**

5. The following diagram shows the production of sex cells and the process of fertilisation in humans. The numbers indicate the number of chromosomes in the cells.

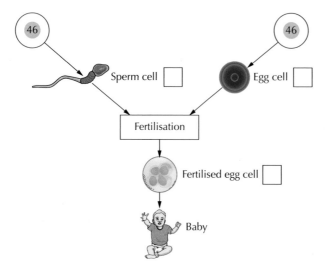

(a) Give the location of sperm production in humans.

_____ **1**

(b) State the term used to describe the chromosome complement of sex cells.

_____ **1**

(c) Copy and complete the diagram above to show the number of chromosomes present in each sex cell and the fertilised egg cell. **1**

(d) State the term used to describe a fertilised egg cell.

_____ **1**

6. In pea plants, tall height (T) is dominant to dwarf height (t). Two pea plants each with the genotype Tt were crossed.

(a) Copy and complete the Punnett square below.

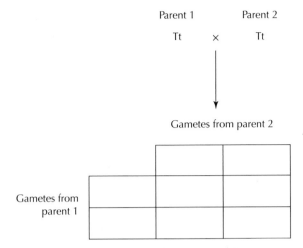

Parent 1   Parent 2

Tt   ×   Tt

Gametes from parent 2

Gametes from parent 1

2

(b) State the expected ratio of tall to dwarf pea plants in the offspring.

_____ tall : _____ dwarf   **1**

(c) The cross actually produced 52 tall plants and 13 dwarf plants. Express this result as a simple whole number ratio.

_____ tall : _____ dwarf   **1**

7. The diagram below shows the human heart.

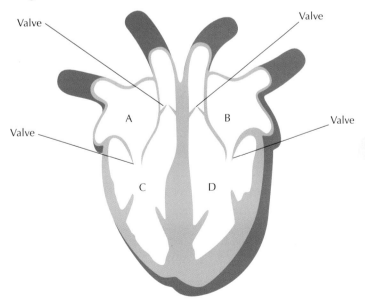

Valve   Valve

A   B

Valve   Valve

C   D

(a) Name chamber C.

_____ **1**

(b) State the function of the heart valves.

_____ **1**

(c) Name the blood vessels that supply the heart muscle cells with oxygen.

_____ **1**

8. There are three different types of blood vessel through which blood travels.
   Choose **two** of these vessels and compare their structure.

_____

_____

_____

_____

_____

_____

_____

_____

_____ **3**

9. A student carried out an investigation into the relationship between running speed and heart rate. She used a treadmill to run at a specific speed for 1 minute and measured her heart rate immediately after. She rested for 2 minutes before running at the next speed.

   The results of the investigation are shown below.

Running speed (km/hour)	Heart rate (beats per minute)
0	70
2	77
4	82
6	95
8	112
10	143
12	175

(a) Why was it good experimental practice to rest for 2 minutes in between each test?

_____ **1**

(b) Draw a **line graph** to show the effect running speed has on heart rate.

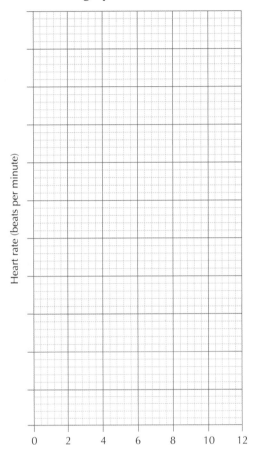

2

(c) Calculate the percentage increase in heart rate as the student's speed increased from 0 km/h to 12 km/h.

_____ % **1**

(d) Describe the relationship between running speed and heart rate.

_____

_____ **1**

(e) Describe how the student could make her results more reliable.

_____ **1**

14   ECOSYSTEMS

Definitions of ecological terms • Food chains • Interactions of organisms in food webs • Niches • Competition in ecosystems • Interspecific competition • Intraspecific competition

15   DISTRIBUTION OF ORGANISMS

Distribution of organisms • Biotic factor – predation • Biotic factor – grazing • Biotic factor – disease • Biotic factor – food availability • Predator–prey interactions • Abiotic factors and the distribution of organisms • Abiotic factors – pH • Abiotic factors – temperature • Measuring abiotic factors • Measuring abiotic factors – light intensity • Using light meters • Sources of error when sampling to measure light intensity • Ways that the errors can be minimised • Measuring abiotic factors – temperature • Using soil thermometers or temperature probes • Sources of error when measuring temperature • Ways that the errors can be minimised • Measuring abiotic factors – pH • Using pH meters • Ways in which the errors can be minimised • Measuring abiotic factors – soil moisture • Using moisture meters • Potential sources of error when measuring moisture • Ways that the sampling errors can be minimised • Sampling • Sampling and adequate replication • Quadrats • Using a quadrat • Limitations and sources of error when sampling using quadrats • Ways that sampling errors can be reduced • Using quadrats to estimate the population size • Pitfall traps • Using a pitfall trap • Limitations and sources of error when sampling using pitfall traps • Ways that errors can be reduced • Need for keys • Paired statement keys • Making keys • Biodiversity • Factors influencing biodiversity • Pollution • Pollution of water • Pollution of air • Habitat destruction • Over-exploitation • Indicator species

16   PHOTOSYNTHESIS

The process of photosynthesis • Photosynthesis experiments • Photosynthesis – Stage 1: Light reactions • Photosynthesis – Stage 2: Carbon fixation • The uses of sugar by plants • Limiting factors

17   ENERGY IN ECOSYSTEMS

The loss of energy from a food chain • Pyramid of numbers • Pyramid of energy • Comparison of pyramids of numbers and energy

18   FOOD PRODUCTION

Increasing human population • Intensive farming • Features of intensive farming methods • Intensively farmed crops • Chemical elements • The importance of nitrates to plants • The importance of nitrates to animals • Fertilisers • Algal blooms • Reducing problems caused by using fertilisers • Genetically modified (GM) rice plants • Pesticides • Pesticides and the food chain • Alternatives to the use of pesticides • Biological control • Need for biological control • Advantages of biological control • Disadvantages of biological control • Biological control in action • Genetically modified (GM) crops • Advantages of GM crops • Disadvantages of GM crops • Bt toxin

19   EVOLUTION OF SPECIES

Mutation • Causes of mutations • Types of mutations • Variation • Adaptations • Adaptations to extreme habitats • Natural selection • Natural selection in action • Peppered moth • Resistant bacteria • Rapid natural selection • Species • Process of speciation

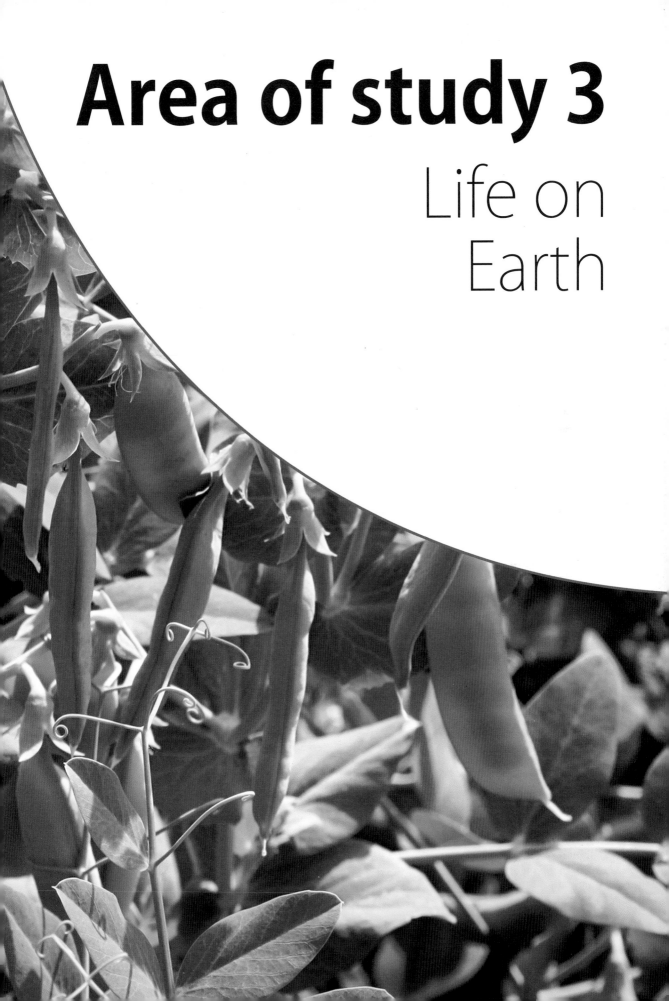

Area of study 3

Life on Earth

14 Ecosystems

Definitions of ecological terms

The correct **definitions** need to be used when studying ecology. Ecology is a branch of biology that focuses on the relationships found in the ecosystems. These relationships can be:

- between different types of organisms or they can be
- between organisms and their physical surroundings.

An **ecosystem** is a biological system. It consists of all the organisms living in the area. All the organisms are called the **community** and the area they live in is called the **habitat**. The organisms present in the habitat will be influenced by their **interactions** with the **non-living components** found in the ecosystem. Non-living components include light intensity, moisture levels and wind speed. A large boulder would also be a non-living component (fig 14.1).

Ecosystem = community + habitat + interactions with non-living components

> ### 🔍 Hint
>
> People often get confused when using the terms population and community. In biology, population is used to refer to all the organisms of the SAME species in the habitat.
>
> Community is used to refer to ALL the living organisms in the habitat.

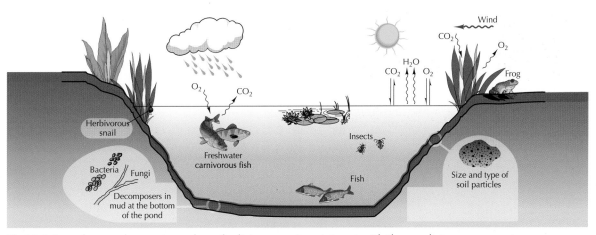

Fig 14.1 *A garden pond ecosystem where the living organisms interact with the non-living components*

Different ecosystems contain a variety of different types of organisms. For example a desert ecosystem contains cacti and camels (fig 14.2). These organisms would not be found in a pond ecosystem, which contains pondweed and sticklebacks (fig 14.3). The variety of different species found in an ecosystem is called **biodiversity**. Biodiversity refers to the number of different types of organisms as well as the relative abundance of each type of organism.

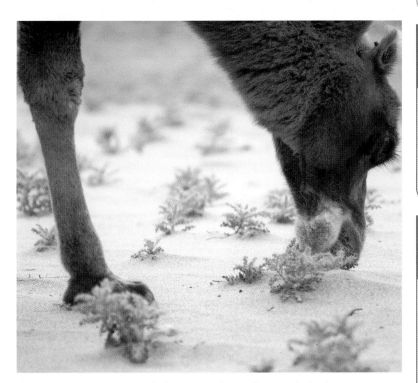

Fig 14.2 *Desert species include cacti and camels. Camels feed on cacti to obtain water*

Fig 14.3 *Pond ecosystems contain pondweed and sticklebacks*

Fig 14.4 *An area of grassland will show biodiversity in the species of plants present*

Fig 14.5 *Grass seed contains several species of grass plants to make sure the lawn will be able to grow vigorously in sun and shade conditions*

Make the link – Social subjects

In geography you may have studied weather and learned about how it can influence the physical environment.

Each different type of organism is a member of one distinct **species**. A species is a group of interbreeding organisms that can produce fertile offspring. If offspring cannot be produced, or if the offspring are infertile, then there is more than one species. For example an area of grassland may contain several different species of grass plant even though the plants look very similar (fig 14.4). Grass seed is sold containing several different species of grass plant (fig 14.5).

Each individual species in the ecosystem makes up a **population**. A population is all the members of one species in the ecosystem (fig 14.6). The **community** is all the species of plants and animals present in the ecosystem (fig 14.7).

Plants in the ecosystem are called **producers**. Produce means to make. Plants are called producers as they can make their own food by the process of photosynthesis. Animals are not able to make their own food. Animals must eat food. Animals are called **consumers** as they eat plants or other animals to obtain food (fig 14.8).

Animals which consume only plants are consumers called **herbivores**. Animals which consume only other animals are called **carnivores**. Animals which consume both plants and other animals are called **omnivores**. Carnivores are animals that hunt and kill other animals. Carnivores are **predators**. The animals that they hunt and kill are their **prey** (fig 14.9).

Fig 14.6 *All ospreys living in Scotland make up the osprey **population***

Fig 14.7 *A **community** of corals*

Fig 14.8 Producers *make food that* **consumers** *eat*

Fig 14.9 *The cheetah is the **predator** of the wildebeest. The wildebeest is the **prey** of the cheetah*

Biology in context

Ecosystems require plants to function. Plants produce food by photosynthesis. The food is eaten by animals. Plants also release oxygen as a by-product of photosynthesis. Oxygen is used by both animals and plants for respiration. Plants provide habitats for other plants as well as animals. Plants can provide shelter from harsh weather conditions for animals.

Most ecosystems require animals in order to function. Prey animals provide food for predators. Animals act as pollinators ensuring that plants can reproduce. Seeds produced by plants are dispersed by animals. Animals can eat fruits and disperse the seeds in their faeces. The fruits can attach to fur and be dispersed. Biodiversity could be reduced without animals. Dominant plants could outcompete other plants. Herbivores feeding on dominant plants allow less dominant plants to grow.

Human activities can influence biodiversity and ecosystems. Sometimes, humans can reduce biodiversity by their activities. For example, people are really concerned about the rate at which rainforests are being cleared. People clear the land so that they can grow cash crops.
However, not all human activities are harmful to biodiversity and ecosystems. There are charitable, and government, organisations which seek to conserve rare and endangered species both within and outside their natural habitats.

The following table gives a summary of ecological definitions

Term	Type of organism	Role in ecosystem
Biodiversity	n/a	The variety of organisms living in the ecosystem.
Habitat	n/a	The place where the organism lives.
Species	Animals and plants	A group of organisms that can interbreed to produce fertile offspring.
Population	Animals or plants	All the organisms of one type living in the habitat.
Community	Animals and plants	All the organisms living in the ecosystem.
Producers	Green plants	Photosynthesise to produce food.
Consumers	Animals	Obtain their energy by eating other organisms.
Herbivores	Animals	Only eat plants.
Omnivores	Animals	Eat both plants and animals.
Carnivores	Animals	Only eat animals.
Prey	Animals	Animals that are eaten by a predator.
Predator	Animals	Animals that hunt and kill other animals.

🔍 Hint

Remember that the arrows in food chains and food webs always show the direction of energy flow. This means that they always point to the feeder – the organism obtaining the energy.

🔍 Hint

The number of individuals in a community will constantly fluctuate. This can happen as a result of disease, predation and competition. If the producer in a food chain is affected by any of these factors this will have a dramatic effect on the organisms further up the food chain.

🌳 Biology in context

Not all food chains and food webs rely on green plants. Green plants need light for photosynthesis. In ecosystems where there is a lack of light, such as in caves, bacteria are the source of food.

🔍 Hint

Remember all food chains and food webs start with plants which produce food which consumers are able to eat.

Food chains

A **food chain** is a simple diagram showing the feeding relationship between organisms. The source of energy entering a food chain is the Sun. This is not normally shown on the diagram. All food chains begin with a green plant called the producer – the source of food. The arrows in a food chain show the direction of energy flow. They always point to the feeder in the relationship (fig 14.10).

In the simple food chain diagram on the next page, the grass is the producer. The grasshoppers are the consumers. They are also herbivores, as they only eat producers. They are the prey of shrews. The shrews are consumers. They are the predators of grasshoppers. Owls are the predators of shrews, which are their prey (fig 14.11).

Food chains are unstable feeding relationships. There is only one source of food for the animals. If one organism is removed then the whole chain fails. They are not common in nature for this reason. In the freshwater food chain (fig 14.12), if the zooplankton were removed then all the water fleas, sticklebacks and pike would die, as they have no alternative food source. The phytoplankton would increase in number, as they are not being eaten.

Interactions of organisms in food webs

Food webs are much more common in nature than food chains. This is because it is rare to find an organism that feeds on only one food source. Food webs are stable feeding relationships because there are alternative sources of food. A food web shows the interconnected food chains that exist in the ecosystem.

Figure 14.13 shows a simplified Antarctic food web. The fish in this food web feed on three different organisms. If the carnivorous plankton were removed, the fish would survive, as they could feed on more krill or herbivorous plankton. This would also mean that the emperor penguins could also survive.

Most food web diagrams are simplified versions of the real-life situation. The food web would become too complicated to read if all the organisms actually in the food web were shown.

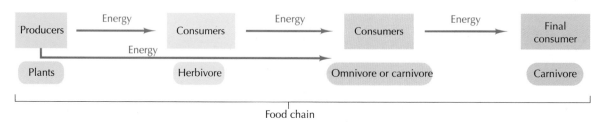

Fig 14.10 *Transfer of energy in an ecosystem*

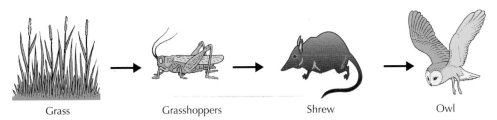

Fig 14.11 *A simple food chain*

Phytoplankton ⟶ Zooplankton ⟶ Water fleas (*Daphnia*) ⟶ Stickleback ⟶ Pike

Fig 14.12 *A simple, unstable freshwater food chain*

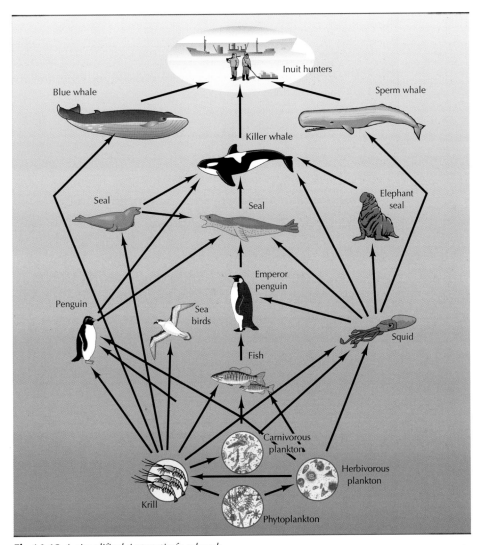

Fig 14.13 *A simplified Antarctic food web*

GO! Activities

Activity 3.14.1 Working individually

1. State what is meant by the term species.

2. State what is meant by the term biodiversity.

3. State what is meant by the term population.

4. Name two producers that start with the letter G.

5. Name two consumers that start with the letter L.

6. Rewrite the following sentence to make it correct.

   Organisms which only eat plants are called (omnivores/ herbivores). These organisms are the (prey/predators) of (carnivores/herbivores).

7. Give three examples of different ecosystems.

8. Give three examples of different habitats found in an oak tree.

9. Ecosystems contain three parts – name these parts.

10. Write down the letters which identify communities from the following list.

    (a) All the slugs on a patio.

    (b) All the robins on a fence.

    (c) All the daisies and dandelions in a grass lawn.

    (d) All the insects living on an oak tree.

    (e) All the mammals living in the savannah.

Activity 3.14.2 Working individually

(a) Copy the table below.

(b) Place examples of the organisms shown in fig 14.13, page 227, into the correct column.

Type of organism		
Producer	Primary consumer	Secondary consumer

Activity 3.14.3 Working in pairs

There are a large variety of different ecosystems on Earth. Some are very large and some are really small. Some ecosystems contain a large biodiversity of plants and animals. Others show a relatively low biodiversity.

Your task is to research different types of ecosystems. You should aim to research at least three different types of ecosystem and present your findings as a PowerPoint presentation.

Your presentation should: contain at least eight slides but no more than twelve; contain the name of the type of ecosystem and where it is found; list at least three animals and three plants found in the ecosystem; give the habitat of the animals you have included; give a description of the average weather conditions found in the ecosystem; state the non-living components of the ecosystem.

Activity 3.14.4 Working in pairs

Read the following sentences and then answer the questions which follow.

'In a freshwater food chain, there are two types of fish – pike and sticklebacks. The sticklebacks are preyed upon by the predator pike. The sticklebacks obtain their energy from eating water fleas (*Daphnia*), which obtain their energy by eating zooplankton. The producers in this food web are phytoplankton.'

- Explain what would happen to the numbers of pike if all the sticklebacks died from an infection.
- Explain what would happen to the zooplankton if all the phytoplankton died due to lack of light.

Rewrite these sentences as a food chain – make sure you include arrows and that the arrows show the direction of energy flow.

(*Continued*)

Activity 3.14.5 Working individually

A food web found in a Scottish forest ecosystem is shown below.

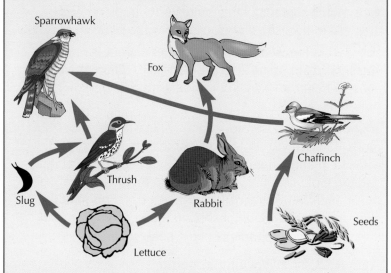

(a) Name the producer(s) in this food web.

(b) Draw a food chain contained in the food web. The chain should consist of four organisms.

(c) Name two prey organisms found in the food web.

(d) Name two predators found in the food web.

(e) Name the organisms the sparrowhawks feed on.

Activity 3.14.6 Working individually

A freshwater ecosystem was studied. A simplified food web, found in this ecosystem, is shown below.

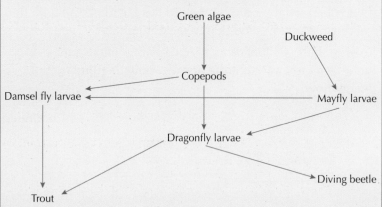

(a) Name the initial source of all the food in this food web.

(b) Name the final consumer(s) in this food web.

(c) Explain what would happen to the number of mayfly larvae if all the duckweed died.

(d) Explain why the numbers of damsel fly larvae could **decrease** if all the copepods died

(e) Explain why the numbers of dragonfly larvae could **decrease** if all the duckweed died.

(f) Explain why the numbers of dragonfly larvae could **stay the same** if all the duckweed died.

(g) Explain the effect on the diving beetles if all the copepods were to die.

I can:

- State that members of the same species are able to interbreed and produce fertile offspring.

- State that biodiversity is a term used to describe the variety that exists within a species and between all living organisms.

- State that a population is all the organisms of one type of species in the ecosystem.

- State that a producer is a green plant that can photosynthesise and produce its own food.

- State that a consumer is an animal that obtains its energy by eating other organisms.

- State that a herbivore is an animal that only eats plants.

- State that a carnivore is an animal that only eats other animals.

- State that an omnivore is an animal that eats both plants and animals.

- State that a predator is an animal that hunts and kills other animals.

- State that prey are animals that are eaten by predators.

- State that a community is all the plants and animals living in an ecosystem.

- State that the habitat is the place where organisms live.

- State that the community and habitat are influenced by non-living (abiotic) components.

- State that an ecosystem is made up of one or several habitats and the community of organisms that live there.

- State that there are many habitats in an ecosystem.

I can:

- State that a food chain is a diagram showing simple feeding relationships between organisms. In nature they are unstable.

- State that a food web is a diagram showing more complex and interconnected feeding relationships between several organisms.

- State that food webs are more common in nature as they are more stable due to the alternative sources of food.

- Describe the effect on a food web if an organism, or organisms, are removed.

Hint

An organism's niche is affected by abiotic factors such as temperature and light. Light is needed for the producers to photosynthesise. Organisms need enzymes to speed up chemical reactions – if the niche is too cold or too hot for the organism's enzymes, then it won't be able to tolerate the temperature in the niche.

Make the link – Biology

Tropical ecosystems have a higher number of producers than other ecosystems. They can therefore support a larger biodiversity than other ecosystems.

Make the link – Social studies

In hotter environments such as rainforests, nutrients are recycled by decomposition much faster than in colder environments like the tundra. How would this affect the biodiversity of plants that grow in these environments?

Niches

The habitat is where the organism lives. The **niche is the role the organism plays within a community**. To describe a niche it is necessary to understand that organisms interact with both the physical and biological parts of the ecosystem.

A niche **relates** to the **resources** the organism **requires** from its **ecosystem**. A niche could be thought of as what the organism does with the resources. It also relates to the organism's **interactions** with the **other organisms in its community**. This could be thought of as how the organism lives. It is important to understand that the organism's niche is not the same as the organism's habitat. The habitat is where the organism lives but the niche is the role that the organism plays in its community.

The niche of an organism depends on how the organism interacts with biotic and abiotic factors.

Interactions in a niche include:

- **Competition** This could be for food.

- **Predation** Predators eat other animals to stay alive.

- **Light** Plants compete for light to photosynthesise.

- **Temperature** Organisms must be able to **tolerate** the abiotic factors in their niche (fig 14.14).

- **Nutrient availability** Decomposition of organic material returns nutrients to the soil.

Biology in context

'The glorious 12th' is a term used to describe the start of the red grouse shooting season on the 12th of August each year. The season lasts until the 10th of December. During this time approximately half a million red grouse will be killed for sport. Large estates use the grouse season to generate money to maintain the estate – people pay to go shooting and the grouse can be sold to restaurants.

To ensure there are enough grouse, gamekeepers burn heather. Burning ensures there is always fresh shoots for the grouse. Wet, boggy land is drained. Gamekeepers also kill predators of red grouse such as stoats, crows and foxes. All these efforts allow more red grouse to survive than the environment could naturally support. This leads to disease spreading within the grouse population. The grouse are given medicines to prevent diseases killing them. The gamekeepers change the role that the grouse play in their niche. Is this a good thing or bad thing?

Make the link – Biology

You will learn more about adaptations that organisms evolve to allow them to live in their environment in Area of study 3, Chapter 19.

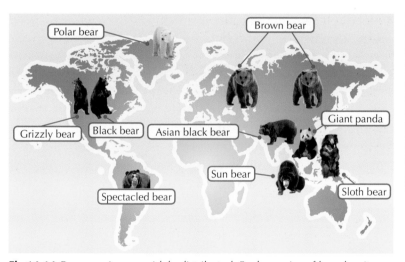

Fig 14.14 *Bear species are widely distributed. Each species of bear has its own niche which varies from the niche of other species of bears*

Make the link – Health and Wellbeing

You may have learned the importance of having a balanced diet that includes all the nutrients needed for growth. Why is it important that ecosystems contain nutrients?

Activities

Activity 3.14.7 Working individually

1. Name two resources that organisms might require from their niche.
2. Name a condition that an organism must be able to tolerate in its niche.
3. Name a resource that an organism may have to compete for in its niche.
4. Rewrite the following sentence to make it correct.

   An organism's (community/niche) is the role that the organism plays in its (community/niche).

(Continued)

Activity 3.14.8 Working individually

The following diagram shows the distribution of two species of barnacles on a rocky seashore in Scotland (fig 14.15).

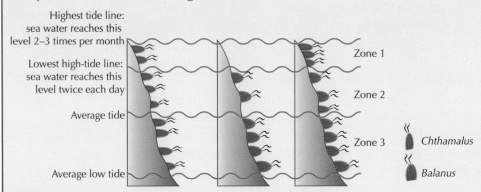

Fig 14.15 *Distribution of barnacles on a rocky seashore in Scotland*

(a) State which zones (habitats) *Chthamalus* can be found living in.

(b) State which zones (habitats) *Balanus* can be found living in.

(c) Suggest a reason for the difference in distribution of these two species of barnacles.

Activity 3.14.9 Working in pairs
Caledonian forests

Scotland was once covered in woodland known as the Caledonian forest. The forest was formed at the end of the last ice age. Less than 1% of the original native forest survives today. This is considered a problem as the forests are the habitat for many rare plants and animals.

(a) On the internet, find the 'Trees for life' website.

(b) Investigate one of the following species using the website:

1. Scottish wildcat
2. Red squirrel
3. Black grouse
4. Scottish crossbill
5. Bracken
6. Atlantic salmon

(c) Produce a slideshow presentation to share with your class. Include in the presentation:

1. The worldwide distribution of the species you have chosen.
2. The distribution of the species in Scotland.
3. The physical and behavioural characteristics of the species.
4. The ecological niche (relationships) these species are involved in.

(Continued)

5. The conservation status of the species, if that is appropriate.

6. Make your presentation 6–8 slides long.

Activity 3.14.10 Working in pairs

12% of Scotland is moorland. Moorland is land that is not farmed or covered in forest. Most of the moorland is covered in heather. Heather is important to the survival of red grouse.

(a) On the internet, go to cairngorms.co.uk/discover-explore/landscapes-scenery/the-moorlands/

(b) Give examples of habitats that are included as moorland.

(c) Give examples of areas in Scotland that are considered to be moorland.

(d) Give reasons why the numbers of grouse in Scotland's moorland may be low.

Activity 3.14.11 Working in pairs

Ptarmigans are birds that are residents of Scotland. They breed in the highest mountains of the Highlands. They have feathered feet.

(a) Perform an internet search for the 'Welcome to Scotland' website. Click on the 'About Scotland' tab and then the 'Scottish Birds' tab. Then click on the Ptarmigan.

(b) State the colour of ptarmigans' feathers in both summer and winter. Explain why the change of feather colour helps these birds to survive.

(c) State how human activities have contributed to the decrease in ptarmigan numbers, especially in winter.

(d) Explain how ptarmigans' feathered feet help these birds to survive in harsh winter conditions.

Activity 3.14.12 Working in pairs

Red deer have been living in Scotland since the last ice age. They are herbivorous animals.

(a) Produce a slideshow presentation including the following points:

1. The habitat that they exploit

2. The numbers that are found in Scotland

3. The lifestyle of red deer

4. Their niche

5. Methods for controlling red deer numbers

(b) The presentation should be between 6 and 8 slides.

(c) The presentation should last 5 minutes.

I can:

- State that a niche is the role an organism plays in its community.
- State that the niche refers to the resources that an organism requires from its ecosystem.
- State that light and nutrient availability are two resources required by organisms in their niche.
- State that the niche also relates to the interactions organisms have with other organisms in their habitat.
- Explain that interactions that organisms have include competition, for example food and also predation.
- Explain that the niche must present conditions that can be tolerated by the organism, e.g. temperature.

Make the link – Social studies

You may have learned about wars that have been fought in the past. Many wars have been fought to acquire more land. Why have so many wars been fought to acquire land?

Hint

Remember that competition occurs when resources which organisms need to survive are in short supply. Competition will be more intense between members of the same species as they require exactly the same resources.

Competition in ecosystems

Animals and plants require particular **resources** to stay alive. Resources needed by plants include light, water, space and soil nutrients. Resources needed by animals include food, water, shelter and mates. If any of the resources needed by animals and plants in the ecosystem are in **short supply** then **competition** may occur.

Interspecific competition

Competition between members of **different species** is called **interspecific** competition. Interspecific competition occurs when different species compete for **one**, or a **few**, of the **resources** they require to stay alive.

This type of competition is usually less intense than other types of competition when resources are in short supply. Different species do not have exactly the same need for resources.

If competition becomes very fierce, one of the species may be forced to move out of the ecosystem. This may happen when resources are extremely scarce. This happened to red squirrels and brown trout.

Red squirrels are native to Britain (fig 14.16). In 2012, there were thought to be around 120000 in Scotland and only 140000 in Britain. The grey squirrel was introduced to Britain (England) in the late 19th century (fig 14.17). They are more efficient feeders than red squirrels and outcompete them. They also carry a virus that kills red squirrels. Red squirrels were forced to move out of their habitats to avoid extinction. They are now only found in areas where there are few grey squirrels.

Brown trout are native to Scotland (fig 14.18). They live in fresh water. Brown trout breed in rivers and migrate to lochs to feed. They return to rivers to breed. In 2007, brown trout were added to the UK Biodiversity Action Plan Priority Species List. One reason for this is interspecific competition from rainbow trout (fig 14.19). Rainbow trout were introduced to Scottish lochs to be farmed. Unfortunately some escaped. They are stronger and more aggressive feeders than brown trout. They outcompeted the brown trout. Brown trout numbers reduced.

Intraspecific competition

Competition between **individuals** of the **same species** is called **intraspecific competition**. **Intraspecific** competition is **more intense** than interspecific competition when resources are in short supply. This is because individuals of the same species require exactly the same resources and have the same adaptations for obtaining these resources. Individuals of the **same** species compete for **all the resources** they require.

Intraspecific competition can lead to the death of the least well-adapted members of the population. They are unable to obtain all the resources they require to survive. This is an example of natural selection in action. Intraspecific competition can regulate the size of the population.

In order to reduce intraspecific competition, some species of birds have evolved territorial behaviour. Robins (fig 14.20) and red grouse (fig 14.21) are examples. A territory is an area that is defended. The territory contains all the resources needed by the bird and its family. The poorer the environmental conditions, the larger the size of the territory. Having a larger territory means a larger area has to be defended. This costs more energy.

Fig 14.16 *Red squirrels are native to Scotland. Their numbers are declining*

Fig 14.17 *Grey squirrels were introduced to Britain. They have outcompeted red squirrels*

Fig 14.18 *Brown trout are native to Scotland. Their numbers are declining*

Fig 14.19 *Rainbow trout were introduced to be farmed. They have outcompeted brown trout*

Fig 14.21 *Male red grouse defend their territory by flying steeply towards the edge of their territory*

Fig 14.20 *Male robins defend their territory by singing a high-pitched song. They also puff out their red breast to appear larger to the intruder*

Make the link – Biology

In Area of study 3, Chapter 19, you will learn about the process of natural selection. Natural selection occurs when there are selection pressures. In intraspecific competition, organisms require exactly the same resources from their environment. Those organisms which are less well adapted to the selection pressures die. Those with favourable alleles survive to reproduce.

 ## Biology in context

Grey squirrels can affect biodiversity by stripping the bark of trees aged 10 – 40 years old. Bark is stripped so the sugary phloem below the bark can be eaten.

Hint

Remember that 'intra' means 'within' and 'inter' means 'between'. Competition within the same species is called 'intraspecific' competition, and competition between different species is called 'interspecific' competition.

 ## Activities

Activity 3.14.13 Working individually

1. Name two different types of competition that occur in ecosystems.
2. State the most intense form of competition.
3. Explain the conditions which exist in an ecosystem which lead to competition between organisms occurring.
4. Explain the differences between each type of competition.

Activity 3.14.14 Working individually

The diagram below shows the distribution of squirrels from 1945 to 2010.

1. Describe the change in distribution of red squirrels in Scotland between 1945 and 2010.
2. Describe the change in distribution of grey squirrels in Scotland between 1945 and 2010.
3. Name the type of competition that occurs between red and grey squirrels.
4. Suggest a reason for the absence of either type of squirrels in the northwest of Scotland.

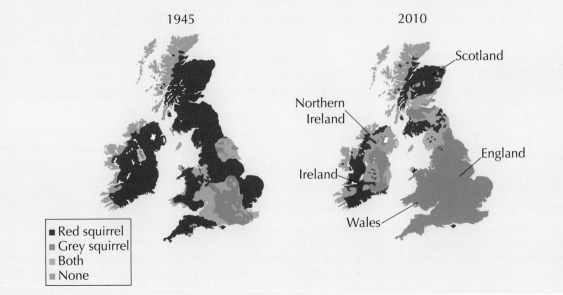

Activity 3.14.15 Working individually

The following graph shows the distribution of red and grey squirrels in Scotland over a 10-year period (fig 14.22).

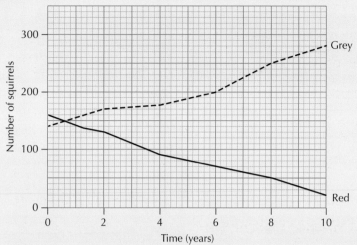

Fig 14.22 *Distribution of red and grey squirrels in Scotland*

1. Identify the time at which the numbers of red and grey squirrels were the same.

2. State the ratio of red squirrels to grey squirrels in year 1 **and** in year 10.

3. Calculate the percentage increase in grey squirrels over the 10-year period.

4. Calculate the percentage decrease in red squirrels over the 10-year period.

5. Predict the number of red squirrels expected in this area if the survey were carried out 5 years after the end of the 10 year period and explain your answer.

Activity 3.14.16 Working in pairs

An investigation into intraspecific competition in mustard seedlings was carried out.

Two yoghurt pots that were wrapped in black paper were collected. 3 g of cotton wool was used in each pot instead of soil. 30 cm^3 of water was added to each pot.

1 g of seeds was planted in one pot; 10 g of seeds was planted in the other. The pots were placed on a warm window ledge for 5 days.

To obtain results, the seedlings were removed from the pots and placed in an oven to dry to a constant mass.

1. Draw a diagram to show the contents of the pots. Make sure you label your diagram.

2. Suggest why the pots were wrapped in black paper.

3. Suggest why cotton wool was used instead of soil as the growing medium.

4. Give two resources the seedlings could be competing for.

5. Explain why the pots were placed on a warm window ledge.

6. Explain why the seeds were allowed to grow for 5 days.

7. Name the pot where intraspecific competition was most intense.

8. Suggest why the seedlings were dried to a constant mass before the results were obtained.

(Continued)

Activity 3.14.17 Working in pairs

An investigation was carried out using barley and pea plants.

Four pots each containing five seeds were planted. The contents of the pots are shown in the table below.

Pot number	Contents of pot	
	Seed planted	Nutrient missing from soil
1	Barley	None
2	Barley	Nitrogen
3	Pea	None
4	Pea	Nitrogen

The seeds were allowed to grow for 4 weeks. The heights of the seedlings were recorded.

Type of plant	Barley		Pea	
Nutrient missing from soil	None	Nitrogen	None	Nitrogen
Average height of seedling (cm)	30.0	24.0	34.0	30.6

(a) State the reason for including pots 1 and 3 in the investigation.

(b) Give two variables not mentioned that need to be controlled to make the experiment valid.

(c) Suggest how the reliability of the results could be improved.

(d) Calculate the percentage decrease in the average height of the barley and pea seedlings when grown without nitrogen.

(e) State which plant grew best when nitrogen was missing from the soil.

I can:

- State that when resources are in short supply in an ecosystem this results in competition between organisms.
- State that interspecific competition occurs between individuals of different species.
- State that intraspecific competition occurs between members of the same species.
- State that interspecific competition is competition for one, or a few, of the resources that the organism requires. Whereas, intraspecific competition is competition for all the resources that an organism needs to survive.
- State that intraspecific competition is more intense than interspecific competition.

15 Distribution of organisms

You should already know:

- Ways to sample living organisms from different habitats.
- How to use branching keys to identify organisms.

Learning intentions

- List biotic factors which affect the distribution of organisms in an ecosystem.
- List abiotic factors which affect the distribution of organisms in an ecosystem.
- Describe methods to sample abiotic factors in a habitat.
- Identify sources of error that might arise when sampling abiotic factors.
- Explain methods that can be used to minimise sources of error when sampling abiotic factors.
- Describe sampling techniques for plants and animals using quadrats and pitfall traps.
- Evaluate limitations and sources of error that might accompany the use of quadrats and pitfall traps.
- Explain methods that can be used to limit the sources of error when using quadrats and pitfall traps.
- State why keys are used to identify organisms.
- Use and construct paired statement keys to identify organisms.
- Explain the effect of biotic and abiotic factors on biodiversity and the distribution of organisms.
- Give some examples of indicator species.
- Explain what indicator species can tell us about environmental quality and pollution levels.

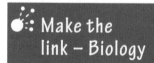 **Make the link – Biology**

In Area of study 3, Chapter 14, you learned about competition between organisms. Competition can affect the distribution of organisms.

Distribution of organisms

The distribution of organisms in an ecosystem is affected by a variety of factors. The distribution of organisms is a way of looking more closely at the biodiversity in the ecosystem. The biodiversity in an ecosystem is affected by a variety of factors. Some of these factors are **biotic factors** and others are called **abiotic factors**. Factors classed as biotic are of living origin. Factors classed as abiotic are of non-living origin.

Biotic factors include **disease, food availability, grazing, predation and competition for resources**. The **abiotic factors** that affect the distribution of organisms in an ecosystem can include **light intensity, pH, temperature and moisture levels**.

Biotic factor – predation

Predation occurs when an animal hunts and kills another animal for food. Predators are carnivores. Predators kill their prey.

Biotic factor – grazing

Grazing is the term used to describe animals feeding on continuously growing grass plants. Over-grazing can occur when the herbivore population is high. Under-grazing can occur when the herbivore population is low.

Moderate grazing can increase biodiversity. Moderate grazing allows the less dominant plants to survive. This is because the reduction in dominant plant species allows the less dominant plant species to obtain more light and soil nutrients (fig 15.1).

Biotic factor – disease

Disease influences the distribution of organisms as if an organism is susceptible to the disease it will die. Individuals which are resistant will survive.

Biotic factor – food availability

Food availability influences the distribution of organisms as if there is not enough food available for all members of the population some individuals will die due to starvation.

Biotic factors have less effect when the population size is small and the habitat contains enough resources to support the population. When the population increases, the effect of the biotic factors increases.

It is important to realise that:

- Single biotic factors may not affect the population on their own. Two or more factors may affect a population at the same time.

Fig 15.1 *Grazing animals can affect plant biodiversity*

🌳 Biology in context

The Great Barrier Reef, off the coast of north eastern Australia, is the largest coral reef in the world. It is the habitat for 400 species of coral and over 1500 species of tropical fish. It is a breeding area for humpback whales and the habitat of endangered species such as green sea turtles. It was listed as a World Heritage Site in 1981.

Worldwide, corals have suffered 'bleaching'. This occurs when the corals get stressed and purge the algae which live inside them, forcing the algae into the sea water. This is a massive problem to the coral as the algae provide most of the energy that keeps the coral alive. Without algae living inside them the coral die. This affects the whole ecosystem. If this continues to happen, already endangered species may become extinct.

What causes the coral to become stressed? Climate change. The increasing sea temperatures cannot be tolerated by coral. Some scientists think that the Great Barrier Reef may die unless we take more action. It is important for ecologists to study abiotic factors to ensure that biodiversity on Earth is maintained.

🔍 Hint

Remember that if food availability in a habitat increases this could result in an increase in the numbers of individuals in the community. Increased numbers make it easier for diseases to spread.

The table below summarises the effect of biotic factors on population.

Effect of biotic factors on population

Biotic factor	Effect on population	Reason
Competition for food	Population decreases	As the population increases, the competition for food increases. Animals starve and may die. To avoid death they may move out of the habitat.
Predation	Population decreases	As the population increases, predation increases. When population size increases there are more prey for predators. More predators move into the area.
Grazing	Population decreases	When the population of plants increases there is more food for herbivores. More herbivores move into the habitat and eat the plants, decreasing the plant population.
Disease	Population decreases	As the population increases, the spread of disease increases. When the population size increases, animals and plants live closer to each other. This makes it easier for disease to spread. More organisms die.

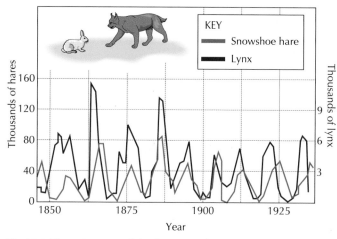

Fig 15.2 *There is a time delay between the increase in hare numbers and the increase in lynx numbers*

> 🌳 **Biology in context**
>
> Gases such as carbon dioxide and water vapour prevent heat from radiating from the Earth's atmosphere into space. This increases the temperature on Earth. This is known as the greenhouse effect.

Predator–prey interactions

The effect of biotic factors on population size can be seen using a classic example of predator–prey interactions. In this case, the snowshoe hare is the prey of the Canadian lynx (fig 15.2). When the hare population increased, there was a small time delay and then the lynx population increased.

However, the increased number of lynx caused the hare population to decrease. There was more predation. Due to the reduced number of lynx, the hare population recovered and their numbers increased. This then caused the predators to increase.

Abiotic factors and the distribution of organisms

Factors that are classed as abiotic are non-living factors. They tend to be physical factors that affect the distribution of organisms in the habitat. Abiotic factors can be measured. Examples of **abiotic factors** include **pH**, **temperature**, **moisture** and **light intensity**.

Abiotic factors – pH

Sulphur dioxide and nitrogen dioxide are released when fossil fuels are burned. These gases dissolve in the water contained in clouds, forming dilute acids. When it rains, these acids fall to Earth as acid rain. Acid rain does not reduce biodiversity directly. Instead, the acid rain damages the leaves of plants. This reduces their ability to photosynthesise.

Abiotic factors – temperature

Climate influences the distribution of organisms. The climate is the pattern of weather conditions experienced in areas over a long period of time. Temperature differences occur due to differences in latitude. Areas around the equator experience higher temperatures compared to those found at the poles which experience colder temperatures.

Make the link – Social studies

You may have studied climate in geography so you will know that the climate on Earth is changing. This is a problem for many species as they may not be adapted to cope with these changes.

Hint

You have learned that herbivorous animals eat only plants in Area of study 3, Chapter 14. Cattle and sheep are examples of animals that graze. Grazing is a biotic factor that affects the biodiversity in an ecosystem.

Activities

Activity 3.15.1 Working individually

1. Give a list of five biotic factors.
2. Give a list of four abiotic factors.
3. Explain what a biotic factor is.
4. Explain what an abiotic factor is.
5. Explain why a population may decrease if it experiences interspecific competition for food from another species within its ecosystem.

Activity 3.15.2 Working individually

The graph below shows the results of an investigation into the relationship between a predator and prey over a period of time (fig 15.3).

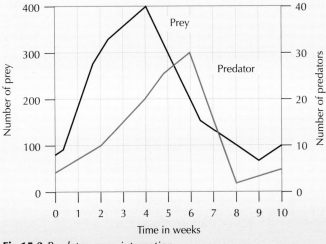

Fig 15.3 *Predator–prey interaction*

(continued)

1. State the number of prey present at the beginning of the investigation.

2. State the number of predators present at the beginning of the investigation.

3. Describe the changes that occur in the number of prey over the 10-week investigation.

4. Describe the changes that occur in the number of predators over the 10-week investigation.

5. State when the highest numbers of prey are present.

6. State when the highest numbers of predators are present.

7. State the number of weeks it takes for the number of predators to reach its maximum after the number of prey reaches the maximum.

8. Explain why there is a time delay before the number of predators reaches the maximum value.

9. Explain why the number of prey decreases from week 4.

10. Explain why the number of predators decreases from week 6.

11. Explain why the number of prey begins to increase from week 9.

Activity 3.15.3 Working individually

The graph shows the biodiversity and biomass of plants in a field over a four-year period. Biomass can be thought of as biological weight.

The data were collected to investigate the effects of grazing on biodiversity.

The field was not grazed in years 1 and 2. In years 3 and 4 sheep were grazing in the field (fig 15.4).

1. Describe the relationship between biomass and biodiversity in years 1 and 2.

2. Describe the effect of the grazing sheep on biomass and biodiversity.

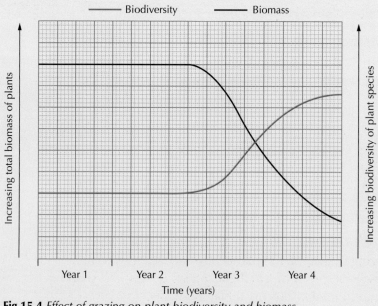

Fig 15.4 *Effect of grazing on plant biodiversity and biomass*

3. Suggest a reason why there was a time delay before the biodiversity increased once sheep began grazing the field.

Activity 3.15.4 Working in groups

Barn owls are nocturnal predators. They feed on small mammals at night. This makes it more difficult to observe them feeding.

Barn owls eat some food items that they cannot digest. This includes the bones of their prey. The undigested material is regurgitated. This material is known as owl pellets.

Search the internet for the Barn Owl Trust website to see an owl-pellet dissection. From this you will be able to learn more about owls as predators. On their homepage search for 'pellet' and then look at the links on the right of the page.

I can:

- State that biotic factors are those that have a living origin.
- State that abiotic factors are those that are non-living and have a physical origin.
- List predation, grazing, disease, food availability and competition for resources as biotic factors that affect the distribution of organisms.
- List pH, temperature, moisture and light intensity as abiotic factors that affect the distribution of organisms.

Measuring abiotic factors

Thinking about experiments 4

An **abiotic factor** is a non-living condition that affects the growth and distribution of organisms in a habitat. Abiotic factors tend to be related to the weather conditions in the habitat.

Measuring abiotic factors – light intensity

Light intensity is an important abiotic factor to sample. Green plants need light to be able to photosynthesise. Plants living in areas with high light intensity are called sun plants. These plants are fast-growing as they receive lots of light.

Shade plants live in areas of low light intensity. This could be in woodland or under the shade of taller sun plants. They are slower-growing plants. Shade plants have adaptations to allow them to survive in areas of low light intensity. Sun and shade plants contain different photosynthetic pigments. Shade plants' leaves tend to be darker green in colour than sun plants.

Fig 15.5 *The light-sensitive layer lies on top of the meter. It is this layer that is pointed towards the area of highest light intensity*

Fig 15.6 *A simple thermometer*

Light intensity also affects the distribution of animals. Animals living in leaf litter move out of light and into shaded conditions. If they are moved into light they move quickly to reach more shaded conditions. This increases the chance of survival.

A light meter is used to measure the light intensity at the soil surface. It has a light sensor to detect light intensity (fig 15.5).

Using light meters

The light-sensitive sensor is pointed at the area of highest light intensity. The reading is allowed to stabilise before the reading is taken.

Sources of error when sampling to measure light intensity

There are potential sources of error when sampling to measure light intensity.

- The reading from the meter may not be accurately read and recorded.

- Light intensity could change during the sampling process.

Ways that the errors can be minimised

The following precautions can be taken:

- Aim to take all the readings at the same time of day. This can be difficult. It is better to take several readings at each sample site and work out the average reading.

- Make sure that the light sensor is not pointing towards the users while the reading is being taken.

Measuring abiotic factors – temperature

Temperature is an important abiotic factor to sample. The enzymes of each plant and animal have an optimum temperature. Metabolism is controlled by enzymes. Organisms therefore require an optimum temperature to survive.

A soil thermometer or temperature probe can be used to measure the soil temperature. Air temperature can be measured using a thermometer (fig 15.6).

Using soil thermometers or temperature probes

The thermometer (fig 15.7) or probe is gently pushed down into the soil. The thermometer or probe should be left in the soil for a few minutes. When the reading remains constant the reading should be taken.

Sources of error when measuring temperature

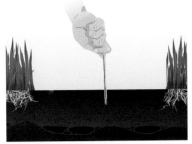

Fig 15.7 *The thermometer is left in the soil until the temperature stabilises*

There are potential sources of error in measuring temperature.

- The thermometer or probe may not be inserted deeply enough into the soil.

- The reading may not measure soil temperature – it may still be reading the air temperature.

- The bulb of the thermometer is being held in the hand.

- The thermometer may be in direct sunlight.

Ways that the errors can be minimised

The following precautions can be taken:

- Try to push the thermometer into the soil so the bulb is covered, or the probe is in the soil to half its depth. If this is not possible, remove and reposition.

- Allow the reading to stabilise before attempting to take a reading. This will usually need a few minutes.

- It is the expansion or contraction of the liquid in the bulb that indicates the temperature. If you hold the bulb it will measure your hand temperature.

- The thermometer must be in shade when measuring air temperature. If it is in direct sunshine the thermometer will give a falsely high reading.

Measuring abiotic factors – pH

pH is an important abiotic factor to sample. pH is a measure of the acidity or alkalinity of the habitat. Acidic conditions destroy soil nutrients and make it harder for plants to absorb water. Some organisms are adapted to live in acidic or alkaline conditions. pH can be measured using a pH probe or pH paper, or by using a chemical test. There are many chemicals that can be used to measure pH.

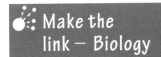

🔍 Hint

Remember that when you are measuring an abiotic factor, you need to give the equipment that you are using time to adjust to the new conditions.

This means that before you take any readings you have to take into account how long it will take for the readings to stabilise.

🧫 Make the link – Biology

In Area of study 1, Chapter 4, you learned about enzymes. Enzymes work best in optimum conditions. Temperature and pH can affect how well enzymes work.

Biologists measure pH to determine if the conditions in the ecosystem are suitable for living organisms.

Using pH meters

The pH meter (fig 15.8) is used in the same way as a moisture meter. There are potential sources of error in measuring pH:

- The reading may be contaminated by soil left on the probe from the previous sample.

- Not enough samples might be taken.

Fig 15.8 *A pH meter can test whether soil is acidic or alkaline*

Ways in which the errors can be minimised

The following precautions can be taken:

- Wiping the probe between each sampling to reduce the risk of cross-contamination.

- Increasing the number of samples that are taken to reduce the effect of sampling errors.

Measuring abiotic factors – soil moisture

Soil moisture content is an important abiotic factor to sample. It measures how much water is in the soil. Most land organisms cannot survive in very wet or very dry conditions. Most plants cannot live in very wet conditions. The water fills up the air spaces in the soil, meaning the plant does not get enough oxygen for respiration. Most plants cannot live in very dry conditions. They wilt and die when water availability is low. Most animals would die of dehydration in very dry conditions.

Make the link – Sciences

In chemistry you may have learned how to measure pH using pH paper or indicators.

The solutions that were tested may have displayed warning labels on the bottles. This is because very strongly acidic and alkaline substances can cause chemical burns.

Make the link – Social studies

You may have learned about weather sampling, which is done to detect trends in climate. What use could ecologists make of the information gathered?

Using moisture meters

The moisture probe (fig 15.9) is inserted into the soil to a depth of 5 cm. The reading is allowed to stabilise and then the reading is taken.

Potential sources of error when measuring moisture

- There may be moisture on the probe from the previous reading.

- Not enough samples are taken.

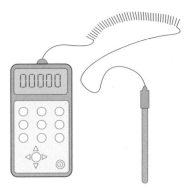

Fig 15.9 *The switch must be switched to moisture when taking a reading if a light-moisture meter is used*

Ways that the sampling errors can be minimised

The following precautions can be taken:

- Wipe the probe with a paper towel before and after taking each reading.

- Take a repeat for each sample site and calculate an average to reduce the effect of sampling errors.

Make the link – Biology

In Chapter 19 of this Area of study (page **321**), you will learn more about adaptations of animals and plants to living in very dry conditions.

Biology in context

The Climate Change (Scotland) Act was passed by the Scottish Parliament in 2009. The Scottish Government recognises that 'climate change will have far-reaching effects on Scotland's economy, its people and its environment . . .'

Climate change is determined by measuring abiotic factors over a long period of time. Ecologists measure abiotic factors and pass their findings to Government. The Scottish Government has acted on the ecologists' findings to make plans for the future. Search the Scottish Government website to find out more information on plans to tackle the anticipated effects of climate change.

Make the link – Health and Wellbeing

Think about when you cook. Accurate measurements need to be made to make sure the product turns out as expected. In PE, accurate measurements of pulse rate need to be taken. This allows you to decide if your fitness is improving. Why are accurate readings important when collecting data?

Hint

When measuring abiotic factors remember the four R's. Readings must be taken:

Randomly to ensure they are **Representative** and they must be **Repeated** to ensure they are **Reliable**.

GO! Activities

Activity 3.15.5 Working individually

1. List four abiotic factors that you could measure.

2. For each of the four abiotic factors listed name the piece of apparatus that you would use to measure them.

3. Give a source of error when measuring each of the four abiotic factors.

4. Suggest a method that could be used to minimise the effect of the source of error for all techniques listed.

Activity 3.15.6 Working individually

A plant found growing next to burns in Scotland is the marsh marigold (fig 15.10).

Quadrats were used to sample the marsh marigold found growing in five different sample sites in the Scottish borders. The quadrats used were 50 cm^2 × 50 cm^2 in size (four quadrats = 1 m^2).

Fig 15.10 *A marsh marigold*

Two abiotic factors were also sampled and averages were calculated for each area.

The results are shown in the table below.

Site of sampling	Average number of marsh marigold in each quadrat	Estimated number of marsh marigold per m^2	Average soil water content (units)	Average soil pH
1	0		10	5.1
2	10		4	6.7
3	15		9	7.3
4	5		3	6.5
5	0		11	5.9

1. State the estimated number of marsh marigold per m² for each of the five sample sites.

2. State the ratio of marsh marigolds found at sample sites 2, 3 and 4.

3. One of the abiotic factors does not appear to influence the numbers of marsh marigold found. Name the factor and explain your reason for choosing this factor.

4. Soil water content was measured using a light moisture meter. Suggest a source of error that could have occurred when measuring the water content and suggest a way that this error could be minimised.

Activity 3.15.7 Working individually

An investigation into the effect of light intensity on the distribution of a type of plant in a garden was carried out. Eight sample sites were investigated. The light intensity and the percentage of the ground covered by the plant were recorded. The results are shown in the table below.

Sample site	Light intensity (lux)	Percentage of ground covered by plant (%)
1	1000	80
2	750	70
3	400	15
4	400	15
5	600	20
6	800	40
7	1900	100
8	1200	90

1. Draw a line graph to show the relationship between sample site and the light intensity **and** the percentage of ground covered by the plant. (Use the same piece of graph paper.)

2. Describe the relationship that exists between the light intensity and the percentage of ground covered by the plant.

3. State the percentage decrease in the percentage of ground covered by plant between sample site 1 and sample site 5.

4. State the percentage increase in the percentage of ground covered by plant between sample site 6 and sample site 7.

5. State the ratio of light intensity found in sample site 1 compared to sample site 2.

6. State the number of times the light intensity in sample site 8 is bigger than the light intensity in sample site 5.

I can:

- State that pH, temperature, soil moisture and light intensity are abiotic factors that can be measured. ⬭ ⬭ ⬭

- Carry out experiments to measure pH, temperature, soil moisture and light intensity. ⬭ ⬭ ⬭

- State that to measure pH, a pH meter is used. To measure temperature, a thermometer is used. To measure soil moisture, a moisture meter is used. To measure light intensity, a light meter is used. ⬭ ⬭ ⬭

- State that a source of error when using a pH meter could be contamination of the probe from a previous sample. ⬭ ⬭ ⬭

- State that a source of error when using a thermometer could be that the thermometer, or soil probe, has not been inserted deeply enough into the soil. ⬭ ⬭ ⬭

- State that a source of error when using a moisture meter could be moisture which has been left on the probe from a previous sample. ⬭ ⬭ ⬭

- State that a source of error when using a light meter could be that the light sensor has been shaded by the user. ⬭ ⬭ ⬭

- State that a way to minimise the source of error when using a pH meter could be to wipe the probe between each reading to reduce the risk of contamination. ⬭ ⬭ ⬭

- State that a way to minimise the source of error when using a thermometer could be to push the thermometer, or soil probe, into the soil to a suitable depth and allowing the reading to stabilise before taking a reading. ⬭ ⬭ ⬭

- State that a way to minimise the source of error when using a soil moisture meter could be to wipe the probe with a paper towel before and after taking each reading. ⬭ ⬭ ⬭

- State that a way to minimise the source of error when using a light meter could be to ensure the sensor is not pointing towards users. ⬭ ⬭ ⬭

Thinking about experiments 5 ⭐

Thinking about experiments 6 ⭐

Sampling

An ecologist may want to **sample** the type of **plants and animals** present in a habitat and to find out the number of different species in a habitat. It would usually be impossible to count every single animal or plant. There could be several reasons for this.

- Number of organisms: There could be too many to count without making an error.

- Time: Large numbers of organisms would take too long to count.

- Habitat damage: Certain counting methods may result in the unacceptable disturbance to organisms.

Sampling allows a **representative** picture of the types and numbers of organisms present to be determined. No sampling method will allow all the different types of animals and plants in the ecosystem to be determined.

Sampling and adequate replication

A sample is a small representative part of the whole. In ecology, samples are taken to try to find out patterns or trends occurring.

To be sure that **representative sampling** is carried out, careful planning is needed. Techniques assume that the samples taken include examples of all organisms present in the habitat. Samples need to be taken using identical sampling apparatus if two different sample areas are being compared. It is also necessary to use the same sampling apparatus to study the same species. This increases the validity of the sampling and allows comparisons to be made.

To be **representative**, several samples have to be taken. The larger the area being studied, the higher the number of samples.

Samples also need to be random to be representative. This means that the experimenter should not choose where the sample is taken.

Quadrats

This technique is used to sample a variety of habitats. It is used to sample plants and animals fixed to the surface of, for example, rocks on a seashore. A **quadrat** is a square frame usually divided into smaller squares (fig 15.11). It is used to estimate a population in a habitat.

Using a quadrat

Quadrats should be dropped randomly. This can be done by the sampler turning away from the area to be sampled and dropping the quadrat over their shoulder. The plants or animals present in the quadrat are then counted.

Limitations and sources of error when sampling using quadrats

There are several limitations and potential sources of error when sampling using quadrats.

Limitations of quadrats:

- Quadrats can only be used to sample plants, and animals which do not move quickly, e.g. barnacles attached to surfaces of rocks.

> ### 🔍 Hint
> Remember that samples are taken when investigating habitats as making accurate counts of the organisms would take too long.

> ### 🔍 Hint
> Remember that replication means that the sampling, or experiment, is repeated. It is good practice to replicate sampling as abiotic factors experienced in the ecosystem will fluctuate.
>
> Replicating the sample gives a more realistic idea of the actual value as averages can be worked out. Averages taken from many replicates minimise the effects of atypical readings.

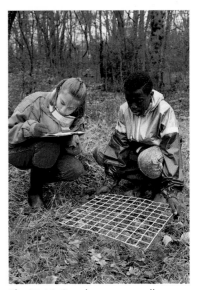

Fig 15.11 *Quadrats are usually subdivided into smaller squares. Only the squares containing the organisms are counted*

255

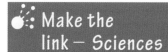

Make the link – Sciences

Experiments can be limited in value due to experimental error caused by the apparatus used.

Scientists try to minimise these errors by using different techniques when they repeat experiments.

Scientists constantly strive to make experimental technique better by thinking...

'What would happen if I changed...'

Hint

Remember, as the size of the area to be sampled increases the number of quadrats used for sampling must also increase to make the results representative.

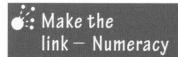

Make the link – Numeracy

In maths you may have learned about measuring areas and volumes. Biologists make use of these relationships when estimating population size.

- The size of the available quadrat may not be suitable for the area, or species, being sampled e.g. a 50 x 50 cm quadrat would be too large to sample algae living on a tree trunk.

Sampling errors when using quadrats:

- Organisms in the quadrat may be wrongly identified.
- Organisms in the quadrat may be wrongly counted.
- Failure to drop the quadrat randomly.
- Too few samples may be taken to be representative. This is especially true when the organisms are clustered in small areas throughout the whole habitat, e.g. plants that are only found in the shade of scattered trees in an area of grassland.

Ways that sampling errors can be reduced

There are ways to reduce sampling errors:

- Select an appropriately sized quadrat for the area being studied.
- Use a key to make sure that the organisms are correctly identified.
- Apply 'rules' of inclusion for counting or not counting, e.g. if organisms are only part in the quadrat, make a rule for counting them. You could only count organisms where at least half is in the quadrat. Another method would be to count only those organisms on the top and right-hand side of the quadrat.
- Use a method to randomise the placing of quadrats, e.g. use dice to provide a pair of random numbers where the quadrat should be placed.
- If it is noticed that organisms are in clusters, increase the number of samples taken or use a transect based method.

Hint

If you have a problem where you are asked to estimate the size of a population in a particular area using quadrats, you must pay close attention to the size of the quadrat used.

You need to work out how many quadrats would be needed to take up an area of 1 m².

Once you know this you can work out how many quadrats you need to cover the entire area.

People often think that there would be two quadrats per metre squared when the quadrats measure 50 cm × 50 cm. There would actually be four quadrats per square metre.

Using quadrats to estimate the population size

The population of an organism can be estimated by applying quadrat results to a whole habitat.

For example:

The area sampled was 25 m long and 25 m wide. The total area was 625 m².

10 quadrats were used in the area. The total number of individuals of a species of plant in the 10 quadrats was 58. This means the average number per quadrat was 5.8.

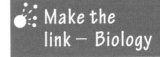

Make the link − Biology

In Area of study 3, Chapter 18, you will learn more about the damage caused to the environment by certain farming techniques. Fertilisers can leach into fresh water and pesticides kill not just pests. Farmers used these chemicals without intending to damage the environment.

If you set up pitfall traps you need to be mindful of the damage you could cause to the environment.

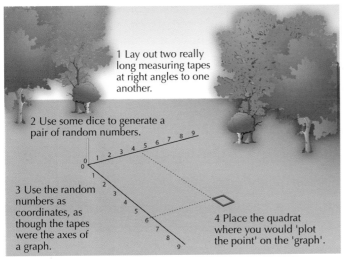

1 Lay out two really long measuring tapes at right angles to one another.

2 Use some dice to generate a pair of random numbers.

3 Use the random numbers as coordinates, as though the tapes were the axes of a graph.

4 Place the quadrat where you would 'plot the point' on the 'graph'.

Fig 15.12 *Follow these instructions to ensure a random quadrat survey is carried out*

Each quadrat was **0.25 m²** so the total population size is 625 × 5.8 × **4** = 14 500 individuals. See the hint box on the bottom of page 256 for further details of how to carry out these calculations.

Figure 15.12 gives more detail of the way a quadrat might be placed randomly.

Pitfall traps

This technique is used to sample ground-living invertebrate animals including nocturnal species living in areas of bare soil or short grass, or even long grass. A **pitfall trap** consists of a jar and a lid. Leaf litter is plant material such as bark, dead leaves, twigs and needles that have fallen on the ground. This is a habitat for many animals, as it is cool and damp and prevents them from drying out.

Hint

Remember when sampling organisms using pitfall traps you have to ensure that the organisms fall into the trap. Small invertebrates, which live in the leaf litter, will be unable to climb over the lip of the trap if it is above the surface of the soil.

🔍 Hint

Remember, as the size of the area to be sampled increases the number of pitfall traps used for sampling must also increase to make the results representative.

Make the link – Biology

In Area of study 3, Chapter 14, you learned some ecological terms and more about food chains and food webs.

Producers will not be trapped in a pitfall trap. This means that there will be no food for herbivores.

It the trap was emptied, and found not to contain herbivores, it is likely the herbivores have been eaten by other animals caught in the trap.

Make the link – Biology

All laboratory-based experiments are repeated to make the results reliable. This should also happen in ecological sampling. Often laboratory-based experiments are repeated quickly after the initial experiment. In ecological sampling the sampling is repeated often on a yearly basis to detect trends.

Using a pitfall trap

A hole is dug in the ground and an empty jam jar or other suitable container sunk into it. Invertebrates moving through the leaf litter fall into the trap and are caught. The steep sides make it impossible for them to climb out. The best traps have a lid, which can be made from large leaves or stones (fig 15.13).

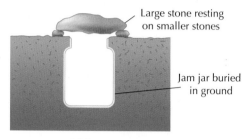

Pitfall trap

Fig 15.13 *The pitfall trap must be level with the soil surface so that animals can fall into the trap*

Limitations and sources of error when sampling using pitfall traps

Some limitations and sources of error are noted below.

- The top of the trap is not level with the soil surface. This means that animals may turn around and not fall into the trap.

- Small winged invertebrates which walked into the trap may be able to escape by flying out of the trap.

- Animals that fall into the trap could be predated if the trap is left set-up overnight and so may not be present when the trap is emptied.

- Animals may die from drowning.

Ways that errors can be reduced

These errors can be reduced in the following ways:

- Dig the hole deep enough and ensure that the top surface of the trap is level with the soil surface.

- Place a lid on top of the trap. This can be done by placing a couple of stones at the edge and resting a large leaf or stone on top. This has to be weighed down by more stones.

- The lid may stop birds from feeding from the trap. Animals that fell into the trap might not be present as they were eaten by carnivores. Empty the trap regularly.

- Place a few small needle holes in the bottom of the trap to allow rainwater to drain away and prevent the animals from dying.

When sampling animals, you should release the animals back into their habitat as soon as you have counted them. This helps to avoid ecological damage caused by the sampling technique.

⚡ Biology in context

Red deer are the largest wild mammals living in Scotland. They are herbivores. Since the extinction of bears, wolves and lynx in Scotland, adult red deer have no natural predators, although the young are sometimes preyed upon by eagles and foxes. The lack of natural predators allowed the population to increase to a number that the habitat could not support. The red deer reduce plant biodiversity. For this reason, the Forestry Commission monitors the numbers of deer. If numbers rise then a planned cull might be licensed.

🔍 Hint

A **cull** is the removal of animals from a group for specific reasons. Culling usually involves killing animals. Red deer are culled to reduce the damage they cause to the habitat.

Activities

Activity 3.15.8 Working individually

1. Name two techniques that can be used to sample plants and animals.
2. For each of the techniques named, list two sources of error which could occur.
3. For each of the sources of error listed, explain how the effect of the source of error could be minimised.
4. Explain the need for adequate replication when sampling plants and animals.
5. Explain steps that could be taken to ensure the sampling technique used is representative.

Activity 3.15.9 Working individually

Quadrats were used to sample the limpets living on a rocky seashore. The quadrats used were 50 cm × 50 cm in size (four quadrats = 1 m²). Three different groups carried out the sampling.

- The positions of the quadrats and the number of limpets found are shown below (fig 15.14).

(Continued)

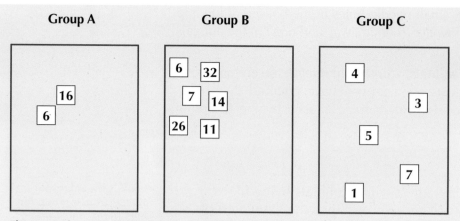

Fig 15.14 *The position of quadrats used by three groups*

1. Copy and complete the table below to show the estimated number of limpets per m².

Group	Average number of limpets in each quadrat	Estimated number of limpets per m²
A	11	44
B		
C		

2. State the sampling error made by group 1.

3. Explain how this error could be minimised.

4. State the sampling error made by group 2.

5. Explain how this error could be minimised.

Activity 3.15.10 Working individually

Pitfall traps were carried out by two different groups to sample the animals living in the leaf litter in the grounds of a school.

Both groups left their traps overnight before collecting their results. The results from both groups are shown below.

Group A	Pitfall trap number	Number of each type of animal found in leaf litter				
		Ants	Spiders	Mites	Woodlouse	Earthworm
	1	3	2	1	0	1
	2	2	1	2	0	0

	Pitfall trap number	Number of each type of animal found in leaf litter				
		Ants	Spiders	Mites	Woodlouse	Earthworm
Group B	1	3	1	1	1	0
	2	3	2	1	1	2
	3	0	2	1	1	1
	4	2	0	3	1	1
	5	2	1	2	2	2

1. State the number of invertebrates found by group A.
2. State the number of invertebrates found by group B.
3. Give the number of times more animals found by group B compared to group A.
4. Explain why group B's results are more reliable than group A's.
5. State the average number of ants found by group B.
6. Give one source of error when setting up pitfall traps.

Activity 3.15.11 Working individually

The internet gives you the opportunity to see how the distribution of a species in an ecosystem changes when using a line transect to sample the biodiversity in an area crossed by a footpath.

Go to saps.org.uk and click on 'Secondary', then 'Teaching resources' and scroll down to 'Ecology Practical 2 – The distribution of species across a footpath'.

I can:

- State that quadrats and pitfall traps can be used to sample animals and plants.
- State that sampling of plants and animals requires several samples to be taken to ensure that the results are representative of the organisms present.
- State that the larger the area being sampled, the higher the number of samples that need to be taken.
- Explain that sampling must be replicated to ensure a reliable picture of the organisms present is obtained.
- Explain that quadrats are used to sample plants or, sometimes, very slow moving animals.
- State that sources of error when using quadrats include placing the quadrat, identifying organisms incorrectly, incorrectly counting organisms and taking too few samples.

(Continued)

- State that these sources of error can be minimised by dropping the quadrat randomly, using a key to identify organisms, making a rule to decide which organisms to include in the counting and taking more samples if the organisms are present in clusters. ⬭ ⬭ ⬭

- Explain that pitfall traps are used to sample animals living in the leaf litter. ⬭ ⬭ ⬭

- State that sources of error when using pitfall traps include the trap not being level with the soil surface, animals escaping from the trap and animals being eaten while in the trap. ⬭ ⬭ ⬭

- State that these sources of error can be minimised by ensuring that the hole is dug deeply enough to house the trap, by placing a lid on the trap to prevent flying insects from escaping and regularly emptying the trap to ensure carnivorous animals don't eat other trapped animals. ⬭ ⬭ ⬭

🔍 Hint

Remember it is not always possible to identify organisms simply by looking at them.

This is because organisms show variations. A biological key takes account of these variations allowing more reliable identification.

Need for keys

When sampling an ecosystem unknown organisms may be collected. It is possible to take the organisms to someone who is likely to be able to identify them. It would be ecologically damaging to the ecosystem if everyone did this. That is one reason why keys are often used.

Keys allow easy identification of unknown organisms.

Fig 15.15 *Bat*	Fig 15.16 *Seagull*	Fig 15.17 *Dolphin*	Fig 15.18 *Brown trout*	Fig 15.19 *Bumblebee*

Statement		
1	Animal has wings Animal doesn't have wings	Go to 2 Go to 4
2	Animal has 1 pair of wings Animal has more than 1 pair of wings	Go to 3 Bumblebee (fig 15.19)
3	Animal has feathers Animal has fur	Seagull (fig 15.16) Bat (fig 15.15)
4	Animal has 2 fins on top of body Animal has 1 fin on top of body	Brown trout (fig 15.18) Dolphin (fig 15.17)

Paired statement keys

A paired statement key is a numbered list of statements. Looking at the organism, decide which of the first two statements fits it best. At the end of each statement, a number signifies which statements to use next. The numbers are followed until the name of the organism is found (figs 15.15–15.19).

The statements are very carefully worded to avoid confusion over characteristics such as colour or size.

Making keys

The diagrams shown in fig 15.20 show the invertebrates collected by pupils when sampling an ecosystem. They are not drawn to scale.

Snail   Spider

Beetle   Woodlouse   Earthworm

Fig 15.20 *Invertebrates found in a sample of an ecosystem*

The invertebrates can be grouped in several ways:

- Invertebrates with legs – spider, beetle and woodlouse. Invertebrates without legs – earthworm and snail.

- Invertebrates with shells – snail. Invertebrates without a shell – earthworm, spider, beetle and woodlouse.

- Invertebrates with spots on body – beetle. Invertebrates without spots – earthworm, snail, spider and woodlouse.

Statement 1	
Invertebrates with legs	Go to 2
Invertebrates without legs	Go to 4
Statement 2	
12 legs or more	**Woodlouse**
Fewer than 12 legs	Go to 3
Statement 3	
Spots on body	**Beetle**
No spots on body	**Spider**
Statement 4	
Shell	**Snail**
No shell	**Earthworm**

These groupings can be used to make a paired statement key (see table on page 263).

A paired statement key can be worked backwards to summarise information concerning organisms.

An earthworm 1 'has no shell' – from statement 4
2 'is an invertebrate without legs' – from statement 1

Make the link – Biology

In Area of study 1 Chapter 1 (pages **14–24**) you learned about the ultrastructure of cells. In Area of study 2 Chapter 7 (pages **92–105**) you learned that tissues are made of different types of cells. You could use a key to identify different types of cells or organelles that are present in cells.

Make the link – Social studies

You may have studied different soil types as an example of a physical feature of a natural environment. How could a key be used to identify different soils?

Hint

Remember, when using paired statement keys, the numbers at the end of the statement indicate which statement to go to next.

GO! Activities

Activity 3.15.12 Working individually

1. State why keys are needed when carrying out sampling.
2. Give a reason why keys are more reliable than pictures at identifying organisms.
3. State what the numbers at the right-hand side of the key show.
4. Give the characteristics of a bat using the key found on page 262.

Activity 3.15.13 Working individually

Keys are not just used in the study of ecology – they are used in all areas of biology. The table below shows some features of five types of white blood cell.

Type of white blood cell	Shape of nucleus	Minimum lifespan of cell	Largest diameter of cell (micrometres)
Neutrophil	Multi-lobed	Hours	12
Eosinophil	Bi-lobed	Days	12
Basophil	Bi-lobed	Hours	15
Lymphocyte	No lobes	Years	15
Macrophage	No lobes	Hours	80

Complete the key using the information given in the table.

1. Nucleus is lobed .... Go to 2
   Nucleus is not lobed .... **(1)**
2. Minimum lifespan is days .... Eosinophil
   (2) ..... **(3)**
3. Largest diameter of cell is 12 micrometres ..... Neutrophil
   Largest diameter of cell is 15 micrometres ...... **(4)**
4. **(5)** .... Lymphocyte
   Minimum lifespan is hours .... **(6)**

State three features of a basophil using the key.

Activity 3.15.14 Working individually

The table below shows some features of British ladybirds.

Name of ladybird	Colour of wing cases	Colour of forebody	Number of spots on each wing case
2-spot	Red	Black	1
7-spot	Red	Black	4
10-spot	Red	White	5
Water	Red	Orange	9
18-spot	Brown	Brown	9

The 7-spot ladybird has one spot that spans each wing case

(a) Construct a paired statement key from the information in the table. The first statement has been completed but you may wish to construct your own first statement.

1. Red wing case .... Go to 2
   Brown wing case 18-spot

(b) State three features of the water ladybird using the key.

I can:

- State that keys are used to identify organisms.
- State that keys are more reliable at identifying organisms than photographs, as organisms show variations.
- State that a paired statement key is a list of numbered statements.

 Biology in context

Biodiversity

Biodiversity is the term used to describe the huge variety seen between organisms of different species. It can also be used to describe the differences between organisms of the same species.

Factors influencing biodiversity

Biodiversity can be influenced by:

- biotic factors
- abiotic factors
- human activities.

Human activities have an impact on biodiversity. In many ecosystems, biodiversity has been reduced due to human activities. These activities include producing pollution, habitat destruction and overfishing.

Pollution

Pollution is something that is added to the environment that causes harm. Pollution of air and water can occur due to human activities.

Pollution of water

Water pollution due to organic waste

Organic waste is material that is added to water that has come from a living source. Examples include blood, sewage and oil. The organic waste provides food for bacteria. The numbers of bacteria increase, and the oxygen concentration decreases as the bacteria use up more oxygen. This reduces biodiversity since larger animals such as fish are unable to survive in low oxygen concentrations.

Water pollution due to oil

Oil leaks from shipping receive lots of media interest. However, oil from cars also causes a lot of pollution. Oil enters the water systems after being washed off the roads by the rain. Oil prevents light reaching plants so their ability to photosynthesise is reduced. This reduces food available to animals, which may then die as a result. Animals can also die from eating the oil. The oil can be caught in feathers making it difficult for birds to fly. This makes them easier for predators to catch.

Pollution of air

In Scotland there is a large biodiversity of lichen species. Lichens are plants. They are actually a fungus and an alga living together. The fungus provides protection for the algae.

The algae provide food for the fungus. They have a symbiotic relationship. There is evidence that lichens have been on Earth for millions of years. They are important in maintaining biodiversity, as birds use them as nesting material, and they provide a home to insects. SNH (Scottish Natural Heritage) states 'Scotland is important for lichens on a European and even global scale'.

Lichens are sensitive to air pollution and are referred to as an indicator species. An indicator species is one that gives information about the levels of pollution in an ecosystem based on its presence or absence. It is sulphur dioxide that the algal part of lichens is sensitive to.

Habitat destruction

Desertification

Desertification is thought to be caused by human activities and the climatic conditions in the area. In certain areas, there are high temperatures for months on end with little rainfall. This leads to drought conditions, making it difficult for crops to grow.

If farming is the main source of income, farmers make as much use of land as possible. This leads farmers to grow as much as possible. Land is not left fallow. Fallow land is land that is rested to allow nutrient levels to recover. The land loses organic matter. This results in reduced vegetation covering the land. This leads to bare soil.

Soil erosion may occur when rain eventually falls. The rainwater washes the soil away. Soil erosion causes many problems. There is less grazing land for animals. Harvests may not give high yields. This results in people struggling to have enough food to eat. Land is therefore left fallow for shorter and shorter periods of time. All of this causes the cycle to start again.

Deforestation

Tropical rainforests are habitats that have been severely damaged by human activities. In many areas, trees in the forests have been removed. The trees may be cut down to be used as building materials or burned as fuel. This contributes to air pollution. The land has then been used for farming: either as arable land or converted to pasture for animals. The land may also be converted for urban use by building homes and new roads.

Tropical rainforests are able to support high biodiversity by providing many habitats and food sources for animals. Removal of the forests has led to the extinction of many species. Estimates suggest that over 100 species are lost every day due to deforestation. Removing trees results in increased competition for food between animals.

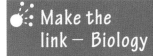

Make the link — Biology

Earlier in this chapter you have learned how to sample abiotic factors which influence the distribution of organisms by learning how to use light moisture meters, pH probes and thermometers. You have also learned how to sample living organisms by using quadrats and pitfall traps.

Make the link — Social studies

Deforestation and desertification are problems in developing countries as people struggle to find land which is fertile enough to grow crops. How could farmers reduce these problems by better land management?

Over-exploitation

Exploitation means to use something in order to gain a benefit. Humans have exploited natural resources for hundreds of years without dramatically altering biodiversity. Problems have occurred due to increases in technology and the increase in the size of the human population. This has led to over-exploitation – taking too much from the environment. This occurred in the fishing industry.

Overfishing

The number of cod found in the Irish Sea and the west coast of Scotland has fallen dramatically due to overfishing. Overfishing has occurred because the catch quota has not been fixed low enough. This resulted in the population not having enough egg producing females to replace the lost fish.

World Oceans Day is an event that hopes to highlight the concern that biologists have regarding plastics ending up in the oceans. People are concerned as animals are being killed by eating large plastics such as plastic bags and food wrappers. Some reports suggest that 100 000 marine animals are killed each year by plastic bags and wrappers. However, it is not just marine animals that are being affected.

It is thought that 'microplastics' are ending up in the food chain when we eat fish that have consumed them. Microplastics are produced in industrial processes and are washed out to sea in waste water. It is these plastics which end up in the human food chain.

GO! Activities

Activity 3.15.15 Working individually

1. List three factors which can affect biodiversity and the distribution of organisms.
2. List two abiotic factors which can affect biodiversity and the distribution of organisms.
3. Rewrite the following sentence to make it correct.

   Biodiversity is the name we give to the (variety/distribution) of organisms found in the (planet/ecosystem). Biodiversity (is constant/changes).
4. List two ways that humans affect biodiversity. For each state whether the human influence increases or decreases biodiversity.

Activity 3.15.16 Working individually

The following graphs show the temperature and precipitation found in tundra in Russia and in rainforest in Belize. The tundra is an area which experiences very little rainfall. It is also a very cold area.

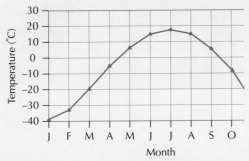

Graph 1 – *Annual temperature*

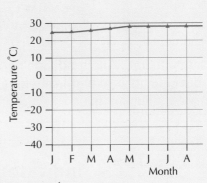

Graph 2 – *Annual temperature*

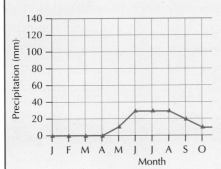

Graph 3 – *Annual precipitation*

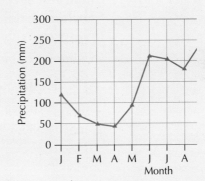

Graph 4 – *Annual precipitation*

1. Use the temperature graphs to identify which of the graphs shows the tundra and which shows the rainforest.
2. Describe the temperature found in each of these ecosystems throughout the year.
3. In which month is there the greatest difference in temperature in the ecosystems?
4. Give the month that a plant living in the tundra may also survive in the rainforest. Explain your reasoning.
5. Use the precipitation graphs to identify which of the graphs is from the tundra and which shows the rainforest.
6. State the month where there is the biggest difference in annual precipitation.

I can:

- List light and moisture as examples of abiotic factors which cause an increase or decrease in biodiversity in an area.

- Describe the effect of human influences that affect environments.

- List air pollution, water pollution and habitat destruction by deforestation, desertification and overfishing as examples of human influences that affect environments.

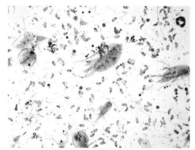

Fig 15.21 *Some freshwater invertebrates found when sampling a pond*

🔍 Hint

Remember pollution is anything added to the environment that causes harm. Industries and power stations could pollute water by releasing clean water that was very hot. The pollution would be thermal pollution. This occurs when the water released is not the same temperature as that taken in. This can lead to stress and the death of organisms.

Indicator species

Indicator species are **species** that indicate **levels of pollution** or **environmental** quality in the environment. They do this by their presence or absence or by their abundance in the environment.

Examples of indicator species are:

- **Freshwater invertebrates:**

  These can indicate levels of pollution in fresh water. This is because they have different levels of sensitivity to dissolved oxygen concentrations. Fresh water can become polluted with organic waste such as sewage. Organic waste acts as a food supply for aerobic bacteria. The bacteria multiply and use up oxygen.

The species of invertebrates living in water depends on the level of pollution. Unpolluted water contains lots of oxygen. Polluted water contains lower levels of oxygen.

If the numbers and species of invertebrates living in fresh water change, this could be an early indication of pollution (fig 15. 21).

Name of indicator species	Level of water pollution	Concentration of oxygen present in the water
Mayfly & stonefly nymphs	Lowest	Highest
Freshwater shrimp & caddis fly larvae	Low or medium	High or medium
Bloodworm & water louse	Medium or high	Medium or low
Rat-tailed maggot & sludgeworm	Highest	Lowest

Fig 15.22 *Lichens are sensitive to air pollution*

- **Lichens:**

    These can indicate levels of air pollution. This is because they have different levels of sensitivity to sulphur dioxide concentration (fig 15.22).

Air can become polluted with sulphur dioxide gas released from factories and power stations when fossil fuels are burned. Lichens are divided into four groups.

1. Crusty lichen – most tolerant to sulphur dioxide pollution

2. Leafy lichen

3. Shrubby lichen

4. 'Hairy' lichen – least tolerant to sulphur dioxide pollution.

In areas that have air pollution only crusty lichens will be found. In areas where the air is unpolluted hairy lichens as well as other types will be found.

Make the link – Biology

In this chapter, you learned how to sample biotic and abiotic factors to show how they affect where organisms live. You now know that organisms can also be sampled to indicate levels of pollution.

Make the link – Mathematics

You have learned how to draw graphs and work out ratios and percentages. In biology, we make use of these skills to work out relationships between organisms and their environments.

Biology in context

Sampling freshwater invertebrates could be damaging to the ecosystem. Removing an organism from its habitat may cause it stress and lead to death. This may be a risk worth taking to ensure other organisms survive. The River Clyde in Glasgow was thought to be heavily polluted due to the long history of industrial use. However, in 2011 SEPA (Scottish Environment Protection Agency) published a report that indicated there are now many species of fish flourishing in the river.

(GO!) Activities

Activity 3.15.17 Working individually

1. Give a definition of the term pollution.

2. Explain what organic pollution is.

3. Give two examples of ecosystems that can be harmed by pollution.

4. Name two different types of indicator species.

5. Explain what is meant by the term indicator species.

6. Name the indicator species that are sensitive to air pollution.

7. Predict the level of pollution in a river which was sampled, and mayfly larvae were found.

8. Predict the oxygen concentration of a river if only rat-tailed maggots were found when the river was sampled.

Activity 3.15.18 Working individually

The following table shows the results from a water-pollution survey carried out on two different streams.

Type of invertebrate found	Number of invertebrates found	
	Stream 1	Stream 2
Mayfly larva	0	8
Caddis fly larva	0	17
Freshwater shrimp	2	78
Water louse	12	36
Bloodworm	40	8
Sludgeworm	96	3

(a) Collect a piece of A5 graph paper. Draw a bar graph of the results in the table. Ensure all the data are on one graph.

(b) State the ratio of sludgeworms in stream 2 compared to those in stream 1.

(c) State the total number of organisms found in stream 1.

(d) State the percentage of sludgeworms in stream 1.

(e) State the total number of organisms found in stream 2.

(f) State the percentage of sludgeworms in stream 2.

(g) State the number of times more water lice were found in stream 2 compared to stream 1.

(h) One stream is more polluted than the other. Suggest which stream was the most polluted. Explain your choice.

(i) Explain why sludgeworms were found in both streams.

Activity 3.15.19 Working in pairs

The following link contains a video that explains the way lichens can be used to indicate air pollution:

http://www.nhm.ac.uk/nature-online/life/plants-fungi/lichens-pollution/index.html

(a) Watch the video.

(b) Produce a summary note you could give a friend who has not viewed the video. The summary note should be 50–100 words in length.

(c) After watching the video you could take part in the OPAL air survey.

I can:

- State that pollution is the addition of substances to the environment that cause harm to organisms. ○ ○ ○

- State that an indicator species tells us something about the level of pollution in the environment. ○ ○ ○

- State that an indicator species tells us something about the quality of the environment. ○ ○ ○

- State that some indicator species give us information by their presence in the environment whereas other indicator species give us information by their absence from the environment. ○ ○ ○

- State that lichens can be used to indicate air pollution. ○ ○ ○

- State that freshwater organisms can indicate organic pollution of freshwater. ○ ○ ○

16 Photosynthesis

The process of photosynthesis

Photosynthesis is the process carried out by all green plants to make food. The word photosynthesis comes from the Greek language for photo, meaning 'light', and synthesis, meaning 'putting together'.

The raw materials – carbon dioxide and water – are used to produce sugar (in the form of glucose) and oxygen. Also essential for the process to take place is light energy (from the Sun) and chlorophyll (the green pigment found in chloroplasts).

Figure 16.1 gives an overview of photosynthesis and the summary word equation.

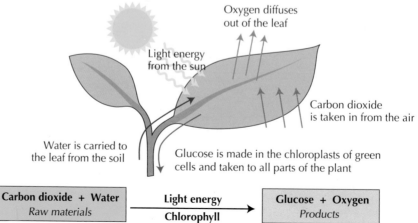

Carbon dioxide + Water	Light energy	Glucose + Oxygen
Raw materials	Chlorophyll	*Products*

Fig 16.1 *Overview of photosynthesis*

Photosynthesis experiments

The sugar produced by plants during photosynthesis is converted to starch.

The following procedure is used to prove that plants store sugar as starch.

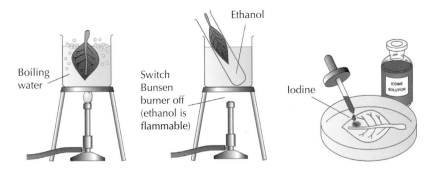

Fig 16.2 *Procedure used to test leaves for starch*

1. A leaf is removed from a plant and boiled in water for a few minutes to soften it.

2. The leaf is then placed in a test tube of ethanol, which is placed into the hot water. The ethanol removes the chlorophyll from the leaf.

3. The leaf is rinsed in water.

4. The leaf is tested for the presence of starch using iodine solution.

5. Iodine solution changes colour from yellow-brown to blue-black in the presence of starch. See figure 16.3.

Fig 16.3 *A leaf showing a positive result to iodine treatment and a leaf showing a negative result to iodine treatment*

The following experiments demonstrate the conditions required by a plant for photosynthesis to take place and the production of oxygen during photosynthesis.

Light is required for photosynthesis

A plant is placed in the dark for 24 hours so that all stores of starch are used up.

Part of a leaf is covered with aluminium foil, and the plant is left in sunlight.

After a few hours the leaf is tested for starch. Only the areas exposed to sunlight are found to contain starch.

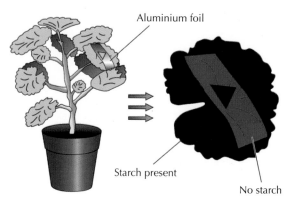

Fig 16.4 *Experiment to show the effect of light on photosynthesis*

☄: Make the link – Biology

Look back to Chapter 6 – Respiration and the experiment using a respirometer (page **73**). Can you make the link between the use of sodium hydroxide in both experiments?

Carbon dioxide is required for photosynthesis

A plant is placed in the dark for 24 hours so that all stores of starch are used up.

The plant is then enclosed in a plastic bag that contains the chemical sodium hydroxide. The sodium hydroxide absorbs the carbon dioxide (fig 16.5). After a few hours the leaf is tested, and no starch is found to be present.

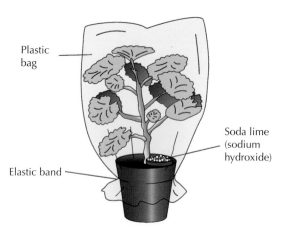

Fig 16.5 *Experiment to show the effect of the absence of carbon dioxide on photosynthesis*

Chlorophyll is required for photosynthesis

A plant with variegated leaves (fig 16.6) is placed in the dark for 24 hours to remove the stored starch. The plant is then placed in sunlight for a few hours. When the leaf is tested for starch, only the green parts of the leaf are found to contain starch (fig 16.7).

Fig 16.6 *Variegated leaf*

Before   After

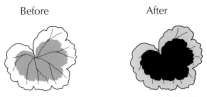

Fig 16.7 *Starch test on variegated leaf*

Oxygen production during photosynthesis

Oxygen production can also be observed in aquatic plants such as Canadian pondweed *(Elodea)*. If the plant is kept in well-lit conditions, bubbles of oxygen are seen to be released from the cut stem of the plant. The gas produced can be collected and tested to prove that it is oxygen (fig 16.8).

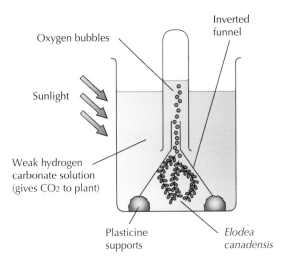

Oxygen bubbles

Inverted funnel

Sunlight

Weak hydrogen carbonate solution (gives CO_2 to plant)

Plasticine supports

Elodea canadensis

Fig 16.8 *Oxygen production by* Elodea canadensis

Photosynthesis is often thought of as a single reaction, but it is in fact a series of **enzyme-controlled** reactions that takes place in a **two-stage process**.

The two stages are:

1. The light reactions
2. Carbon fixation

⚛ Make the link – Biology

Enzymes are biological catalysts (Chapter 4). They are required for photosynthesis. Photosynthesis is carried out by producers and provides the energy for food chains/food webs (Area of study 3 page **226**).

🌳 Biology in context

Using immobilised algae is an interesting way to investigate photosynthesis. Algal balls can be made from a mixture of algae and sodium alginate (a gum extracted from brown algae). They will carry out photosynthesis under the right conditions.

The algal balls are placed into vials containing water and bicarbonate indicator. The vials are then placed under various light intensities ranging from darkness to bright light, along with a control where no indicator is used.

The bicarbonate indicator solution changes colour as the pH changes due to changes in carbon dioxide concentration during photosynthesis. If the concentration of carbon dioxide increases above 0.04%, the solution changes colour from red to yellow. If the carbon dioxide concentration decreases below 0.04%, it changes colour from red to purple. The colour changes allow deductions to be made about how light levels influence photosynthesis (fig 16.9).

Fig 16.9 *Experiment using immobilised algae*

Photosynthesis – Stage 1: Light reactions

The light reactions take place in the first stage of photosynthesis and occur inside chloroplasts. Chloroplasts are organelles found inside plant cells and contain the pigment chlorophyll.

The number of chloroplasts found in a plant varies depending on the type of plant and the number of leaves.

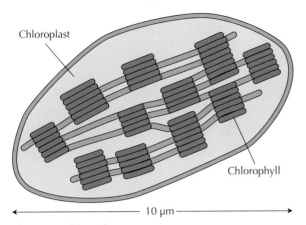

Fig 16.10 *Chloroplast structure*

The diagram of the chloroplast shows where the chlorophyll is found in the chloroplast (fig 16.10).

Light energy from the sun is trapped by **chlorophyll** in the **chloroplasts** and is converted to chemical energy. The chemical energy is used to generate ATP and split water.

Water is split to produce **hydrogen** and **oxygen**, a process also known as photolysis: photo meaning 'light' and lysis 'to split'. The splitting of water releases hydrogen, which is transferred to the carbon fixation stage of photosynthesis. The excess oxygen diffuses out of the cell and is released into the atmosphere.

The light reaction is summarised in figure 16.11.

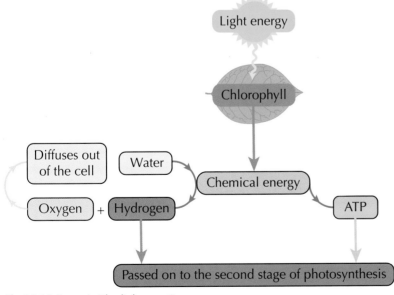

Fig 16.11 *Stage 1: The light reactions*

ATP is required to provide energy in carbon fixation – the second stage of photosynthesis.

Photosynthesis – Stage 2: Carbon fixation

Carbon fixation is the second stage of photosynthesis, and it also takes place inside the chloroplast. The purpose of carbon fixation is to form **sugar**.

Carbon dioxide (from the air) is required for the formation of sugar. Carbon dioxide enters the leaf through pores called stomata (fig 16.12) that are found on the lower leaf surface. These pores open during the day allowing carbon dioxide to enter. Light energy from the sun allows the light reactions to take place, releasing hydrogen from water and generating ATP.

During carbon fixation the energy from **ATP** is used to combine **hydrogen** and **carbon dioxide.** Following a series of enzyme-controlled reactions, this produces the sugar glucose (fig 16.13).

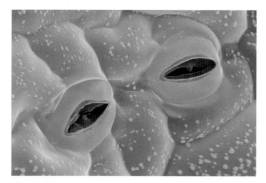

Fig 16.12 *The stomata on the underside of a leaf are visible when viewed under a high-powered microscope*

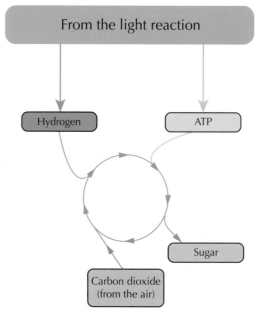

Fig 16.13 *Carbon fixation*

The summary diagram in figure 16.14 shows how the products of the light reactions are used in carbon fixation to produce sugar.

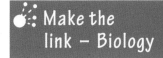

Make the link – Biology

Water is also lost through evaporation when the stomatal pores are open (Area of study 2 page **260**).

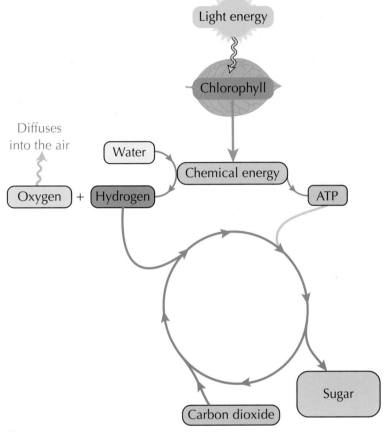

Fig 16.14 *Summary of stage 1 and stage 2*

GO! Activities

Activity 3.16.1 Working individually

1. Write down full answers to the following questions:

   (a) Name the raw materials required for photosynthesis to take place.
   (b) Name the pigment that traps light energy from the sun.
   (c) Give the summary word equation for photosynthesis.
   (d) Name the two stages that make up the process of photosynthesis.

2. Produce a leaflet giving detailed instructions of the procedure used to test leaves for starch. Include pictures.

Activity 3.16.2 Working individually

Write down full answers to the following questions:

1. Name the form of energy that light energy is converted to. Describe what it is used to do.
2. State what happens to excess oxygen produced during photosynthesis.

3. Name the products of the light reactions that are used in carbon fixation.

4. Describe what happens in the process of carbon fixation.

Activity 3.16.3 Working in pairs

Produce a poster to summarise the main two stages in photosynthesis.

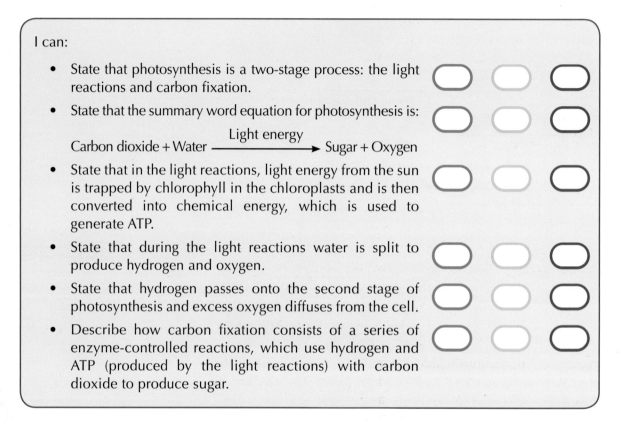

I can:

- State that photosynthesis is a two-stage process: the light reactions and carbon fixation.

- State that the summary word equation for photosynthesis is:

$$\text{Carbon dioxide} + \text{Water} \xrightarrow{\text{Light energy}} \text{Sugar} + \text{Oxygen}$$

- State that in the light reactions, light energy from the sun is trapped by chlorophyll in the chloroplasts and is then converted into chemical energy, which is used to generate ATP.

- State that during the light reactions water is split to produce hydrogen and oxygen.

- State that hydrogen passes onto the second stage of photosynthesis and excess oxygen diffuses from the cell.

- Describe how carbon fixation consists of a series of enzyme-controlled reactions, which use hydrogen and ATP (produced by the light reactions) with carbon dioxide to produce sugar.

The uses of sugar by plants

The sugar made in the process of photosynthesis is a source of chemical energy. This energy is available for **respiration** or can be converted into plant products such as **starch (storage)** and **cellulose (structural)**.

The uses of sugar by plants is summarised in the diagram below (fig 16.15).

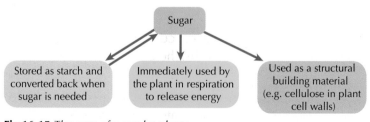

Fig 16.15 *The uses of sugar by plants*

Unlike humans, plants cannot source different food types. The only food type available to plants is the carbohydrate (sugar) made during photosynthesis. The **carbohydrate** made by plants is then used to produce **fats** and **proteins**.

 Activity

Activity 3.16.4 Working individually
Produce a poster or leaflet to show the fates of the sugar produced during photosynthesis.

I can:

- State that the chemical energy in sugar is available for respiration or can be converted into other substances, such as starch (storage) and cellulose (structural).

Limiting factors

When the sugar made during photosynthesis is used by the plant for respiration, this releases energy for growth. The use of sugar for growth allows the plant to grow new parts. The growth of new leaves allows the plant to carry out more photosynthesis and, in turn, make more food.

Factors such as **carbon dioxide concentration** and **light intensity** are essential for photosynthesis to take place. **Temperature** is also important for photosynthesis, as the process is controlled by enzymes, and enzymes are dependent upon temperature.

If any of these factors are in short supply, this will limit the rate of photosynthesis. For this reason, carbon dioxide concentration, light intensity and temperature are called **limiting factors**.

To find out how limiting factors affect photosynthesis, the rate of photosynthesis needs to be measured. This can be done by measuring one of the following over a set period of time: the volume of oxygen released by a plant; the carbon dioxide used by a plant; or the gain in dry mass of a plant.

Aquatic plants like *Elodea canadensis* (Canadian pondweed) and *Cabomba* are commonly used for measuring the rate of photosynthesis. If the plant stem is cut under water, the number of bubbles of oxygen released by the plant can be counted, or collected and measured (fig 16.16).

The experiment shown in figure 16.17 is set up to investigate the effect of light intensity on the rate of photosynthesis.

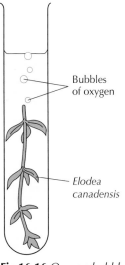

Bubbles of oxygen

Elodea canadensis

Fig 16.16 *Oxygen bubbles released by* Elodea canadensis

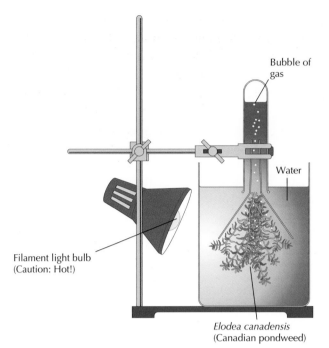

Fig **16.17** *Investigating light intensity*

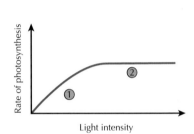

Fig **16.18** *Effect of increased light intensity on the rate of photosynthesis*

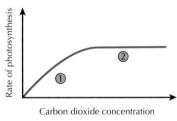

Fig **16.19** *Effect of carbon dioxide concentration on photosynthesis*

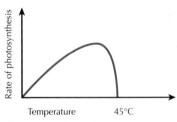

Fig **16.20** *Effect of temperature on photosynthesis*

The distance of the plant to the lamp can be changed to vary the light intensity available to the plant for photosynthesis.

The graph in figure 16.18 shows the effect of increased light intensity on the rate of photosynthesis.

At point 1 on the graph, light intensity is a limiting factor. As the light intensity increases, so too does the rate of photosynthesis. At point 2, the light intensity has continued to be increased, but the rate of photosynthesis has not increased. Therefore, one of the other factors must be in short supply and is limiting the rate of photosynthesis.

The graph in figure 16.19 shows the effect of increased carbon dioxide concentration on the rate of photosynthesis.

At point 1 on the graph, carbon dioxide concentration is increasing, as is the rate of photosynthesis. This tells us that carbon dioxide concentration is the limiting factor. At point 2, the rate of photosynthesis stays the same despite the carbon dioxide concentration being increased. As the plant is being given more carbon dioxide, but the rate of photosynthesis is not increasing, the carbon dioxide concentration cannot be the limiting factor. One of the other factors must be in short supply.

The graph in figure 16.20 shows the effect of increased temperature on the rate of photosynthesis.

At low temperatures the rate of photosynthesis is slow. As temperature is increased, the rate of photosynthesis increases. The plant enzymes involved in photosynthesis work best between 20 and 25°C. As the temperature increases, the rate of photosynthesis begins to slow again, until 45°C where enzymes are denatured, and the rate of photosynthesis drops to zero.

Different factors can affect the rate of photosynthesis at different times. This is shown on the graph in figure 16.21.

Make the link – Biology

Look back to Chapter 4 (Proteins) for a reminder of enzymes that are affected by temperature.

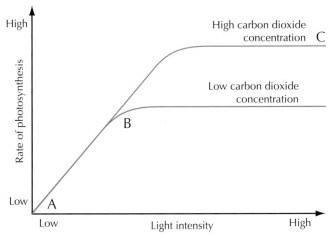

Fig 16.21 *Effects of temperature on photosynthesis*

Between points A and B on the graph (fig 16.21), light intensity is the limiting factor. Increasing the light intensity increases the rate of photosynthesis. After point B the rate of photosynthesis levels off, even though light intensity continues to increase. The limiting factor is no longer light intensity. Increasing the carbon dioxide concentration at point B causes an increase in the rate of photosynthesis, so it is concluded that carbon dioxide concentration is the limiting factor at point B.

At point C the rate of photosynthesis has again levelled off in spite of an increase in both light intensity and carbon dioxide concentration. The limiting factor must be temperature.

Only by eliminating the limiting factor can the rate of photosynthesis be increased. The more photosynthesis that takes place, the more sugar is produced. The more sugar there is, the more energy available to the plant for growth.

GO! **Activity**

Activity 3.16.5 Working individually

Read the information carefully, and then complete the tasks below.

In an experiment carried out to investigate the rate of oxygen production as light intensity is increased, the following results were obtained:

Light intensity (units)	Number of bubbles of oxygen produced in 2 minutes
0	0
5	12
10	21
15	34
20	45
25	45

1. Draw a line graph of the results.

2. Answer the following questions:

   (a) When light intensity is zero, why are no bubbles of oxygen produced?

   (b) Explain why the number of bubbles of oxygen produced can be used as a measure of the rate of photosynthesis.

   (c) When light intensity is increased from 20 to 25 units there is no increase in the number of bubbles of oxygen produced. Why is this?

   (d) What other factors could have had an effect on the number of bubbles of oxygen produced?

   (e) What could be done to improve the reliability of the results?

I can:

- State that limiting factors in photosynthesis are: carbon dioxide concentration, light intensity and temperature.

- Explain that the absence or reduction of any one of the limiting factors reduces the rate of photosynthesis, slowing plant growth.

- Analyse graphs on limiting factors to identify the factors that are limiting photosynthesis at set points.

17 Energy in ecosystems

You should already know:

- Plant and animal species depend on each other for survival.
- Energy flows through food chains.
- Energy can be lost as energy is transferred from one level to the next in a food chain.

Learning intentions

- Explain that energy is lost during transfers of energy from one level in a food chain to the next level.
- Give examples of ways in which energy is lost in energy transfers.
- State the use organisms make of the small quantity of energy that is passed to the next level.
- Give definitions of pyramids of numbers and pyramids of energy.
- Compare and contrast pyramids of numbers and pyramids of energy.
- Give the units of energy used to compare energy present in pyramids of energy.
- Describe why irregular pyramids of numbers occur.

✺: Make the link – Biology

In Area of study 3, Chapter 14, you learned the ecological terms needed to describe energy flow through food chains and food webs. All food chains and food webs start with a producer. They contain consumers – herbivores and carnivores are always present. Short food chains may not contain omnivores.

The loss of energy from a food chain

Energy is lost as it is **transferred** through a food chain. In transfers from one level to the **next** the **majority** of energy is **lost**.

Energy can be lost as **heat**, during **movement** or as **undigested materials**. Undigested materials include bones, hair and fur. Only a **very small** quantity of the energy transferred is used by organisms. The transferred energy is used for **growth**. The energy that is used for growth is the energy that will be available to organisms in the **next level** of the food chain. The ways that energy is used by organisms are shown in figure 17.1.

Energy losses from plants can include light that is not used in photosynthesis. Some of the light hitting plants can pass straight through the leaf or bounce off the surface of the leaf. Of the light energy that is used in photosynthesis, some will be lost as heat from both photosynthesis and respiration reactions.

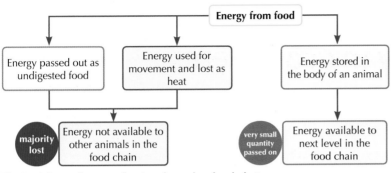

Fig 17.1 *Fate of energy flowing through a food chain*

Energy losses from animals include energy that is lost as heat through respiration. The majority of the energy taken in by animals is lost as heat, faeces and urine, and during movement.

Biology in context

A large quantity of food eaten by cows passes through their digestive system and is lost as undigested materials. This is a waste of energy. For this reason, farmers have traditionally used cow manure as fertilisers.

To reduce climate change, governments are looking for ways to produce renewable energy. In order to produce renewable energy, energy companies are investigating ways to make use of the energy lost in cow manure as well as other green wastes, such as silage from agriculture. Biogas plants are becoming increasingly common.

A biogas digester is filled with manure and silage. Bacteria respire these substances and produce gas. These gases can be used to turn turbines which, in turn, produce renewable electricity.

GO! Activities

Activity 3.17.1 Working individually

1. State what is shown by arrows in food chains and food webs.

2. State three ways in which the majority of energy is lost as energy is transferred from one level to the next in a food chain.

3. Explain how energy which is transferred to the next level in a food chain is used.

Activity 3.17.2 Working individually

Water fleas are the food for sticklebacks but they feed on zooplankton. Zooplankton feed on phytoplankton, which

Make the link – Biology

In Area of study 3, Chapter 16, you found out that photosynthesis is vital for humans to survive. Our food comes either directly or indirectly from the process of photosynthesis. Vegans eat food only from plant sources. Vegetarians eat dairy products and so rely on animals for food as well as plants.

Make the link – Sciences

The law of conservation of energy states that energy can be neither created nor destroyed but can be converted from one form into another. What energy conversion takes place in plants?

Hint

Remember that heat is lost at each link in a food chain.

All organisms produce heat when they respire. Refer to Area of study 1, Chapter 6, to read more about respiration.

(Continued)

are the producers. The last animal in the food chain is pike. Pike feed on sticklebacks.

1. Create a simple food chain using the information above.

2. Give two processes that occur in phytoplankton and pike which result in energy loss.

3. Give two examples of undigested materials created by sticklebacks that cannot be passed on to pike.

4. Zooplankton are tiny animals. List ways in which they might use energy transferred on to them by phytoplankton.

I can:

• State that as energy is transferred along food chains, the majority of energy is lost before it is able to be transferred to the next level of the food chain.

• State that energy can be lost as heat, during movement and in undigested materials such as fur and bones.

• State that only a very small quantity of energy, that has been transferred, is available to be used for growth by the consumer.

• State that only the energy that has been used for growth will be available to the next level in the food chain.

Pyramid of numbers

A food chain shows the feeding relationship between organisms in an ecosystem. It does not tell us anything about the numbers of each of the organisms present at each feeding level. A **pyramid of numbers** is a diagram that shows the **numbers of organisms present at each level of the food chain**.

The producers are always at the base of the pyramid. The final consumer is at the tip of the pyramid. The pyramid shape is due to the decrease in numbers at each link of the food chain. The longer the bar, the higher the number of organisms. In regularly shaped pyramids, the size of the organism increases, moving along the food chain.

The following food chain (fig 17.2) can be represented as a pyramid of numbers (fig 17.3). There are most grass plants and fewest owls so a pyramid shape occurs.

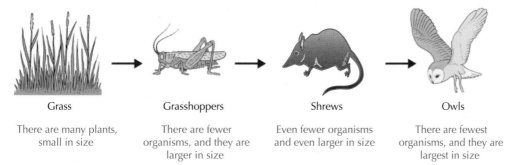

Grass	Grasshoppers	Shrews	Owls
There are many plants, small in size	There are fewer organisms, and they are larger in size	Even fewer organisms and even larger in size	There are fewest organisms, and they are largest in size

Fig 17.2 *A forest food chain*

There are some exceptions to the classic pyramid shape. These pyramids are said to be **irregular**. Some food chains begin with a single large producer, for example a tree. In this case the base of the pyramid (producer) is narrower than that of the primary consumers (fig 17.4).

Another exception is when parasites are feeding on the final consumer at the end of the food chain. In the food chain

lettuce → rabbit → fox

the fox may have parasites called fleas. The fleas are tiny compared to the fox. The pyramid of numbers when the fleas are included would look like this (fig 17.5).

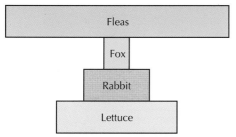

Fig 17.5 *The fleas have a much longer bar than the fox as there are many of them*

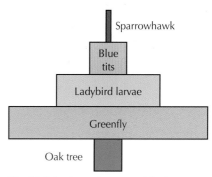

Fig 17.3 *A pyramid of numbers*

Fig 17.4 *An irregular pyramid, since the producer is a single large organism*

Pyramid of energy

A **pyramid of energy** is the most reliable way of representing feeding relationships in a food chain. This is because energy is always lost at each link in a food chain.

A pyramid of energy is a diagram which shows **the total energy contained within organisms present at each level of the food chain**. The units used for energy are **kJ/m²/year**.

Hint

Remember that pyramids of numbers are the only type of pyramids that can have an irregular shape. This is caused when there is one, single large producer at the start of the food chain or when parasites are included in the food chain.

Pyramids of energy are the most difficult to research. This is because the energy content of each population has to be calculated.

Comparison of pyramids of numbers and energy

Zooplankton are animals that feed on phytoplankton. Zooplankton eat phytoplankton very quickly. This means the phytoplankton never have higher numbers than the zooplankton (fig 17.6). However, there is always enough phytoplankton for the zooplankton to feed on. Phytoplankton reproduces very quickly.

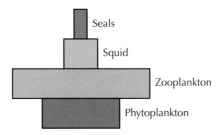

Seals
Squid
Zooplankton
Phytoplankton

Fig 17.6 *An irregular pyramid of numbers. Phytoplankton reproduce very quickly so they are able to support the zooplankton feeding on them*

A pyramid of energy can be drawn for the irregular pyramid of numbers shown by figure 17.6. Energy is lost as heat at each link. This results in the shape always being a true pyramid (fig 17.7).

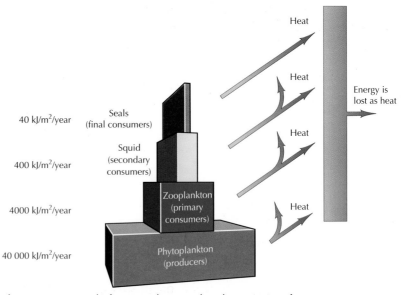

40 kJ/m²/year — Seals (final consumers)
400 kJ/m²/year — Squid (secondary consumers)
4000 kJ/m²/year — Zooplankton (primary consumers)
40 000 kJ/m²/year — Phytoplankton (producers)

Heat
Heat
Heat
Heat
Energy is lost as heat

Fig 17.7 *A pyramid of energy showing that the majority of energy at each level in the food chain is lost as heat*

GO! Activities

Activity 3.17.3 Working individually

1. State what a pyramid of numbers shows.

2. Explain what happens to the number of organisms present when moving along a food chain.

3. State what happens to the size of organisms when moving along a food chain. Explain why this happens.

4. State the term used to describe a pyramid of numbers that does not follow this pattern.

5. Explain why some pyramids of numbers do not follow this pattern.

6. State what a pyramid of energy shows.

7. State the units that should be used for measuring energy when constructing a pyramid of energy.

8. Explain why a pyramid of energy is the best way to represent a food chain.

Activity 3.17.4 Working individually

In a garden, one rose bush provided the food for 500 greenfly. The greenfly were preyed upon by 10 ladybirds, which in turn were preyed upon by one blackbird.

1. Construct a food chain from the information provided.

2. Draw a pyramid of numbers using the information given. State whether the pyramid is a regular or irregular pyramid. Explain how you reached your conclusion (hint – consider the numbers of organisms at each link in the food chain).

3. The rose bush was found to contain 9000 kJ/m²/year of energy at the beginning of the growing season. The majority of energy is lost as energy is passed on to the following level in the food chain. In this food chain it was estimated that 90% of energy was lost from each level before it could be passed on.

   State how much energy would be available to the blackbird.

I can:

- State that a pyramid of numbers is a diagram that shows the number of organisms present at each level in a food chain.

- State that the number of organisms present usually decreases when moving along a food chain.

- State that the size of the organisms usually increases when moving along a food chain.

- State that in some cases irregular pyramids of numbers are formed.
- State that irregular pyramids of numbers can occur when there is a single, large producer at the start of the food chain.
- State that irregular pyramids of numbers occur when parasites are included in the food chain.
- State that irregular pyramids of numbers occur due to different body sizes of organisms in the food chain.
- State that irregular pyramids of numbers can be represented as true pyramids of energy.
- State that a pyramid of energy shows the energy present in the organisms at each level in the food chain.
- State that the units for energy in a pyramid of energy are kJ/m^2/year.
- State that a pyramid of energy always takes a regular pyramid shape.
- State that energy is lost at each link in a food chain.

18 Food production

You should already know:

- Nitrogen in the form of nitrates is needed for plant growth.
- Fertilisers supply nitrates to the soil to help plants grow.
- Fertilisers can be washed into fresh water and increase algal blooms.

Learning intentions

- Describe the effect of increasing population on the need to increase food yield.
- Give examples of chemicals used in farming to increase food production.
- State why fertilisers are used in food production.
- State why pesticides are used in food production.
- Explain why fertilisers contain nitrates.
- Explain why plants need nitrates.
- Explain why animals consume plants.
- State what an algal bloom is and explain why algal blooms occur.
- Describe how algal blooms kill aquatic plants.
- Explain how algal blooms bring about an increase in the number of bacteria in water.
- Explain how algal blooms contribute to a reduction in the oxygen content of water.
- Explain what is meant by bioaccumulation and how it occurs.
- Describe the use of biological control and genetically modified (GM) crops as alternatives to the use of fertilisers and pesticides.

Increasing human population

Make the link – Social studies

In geography, you may have studied human population pyramids. You will know that developed countries have an ageing population.
Developing countries have rapidly expanding populations due to the high numbers of births.

The total human population has grown significantly since the 1850s (fig 18.1). The massive increase in population is called a population explosion. The main increase has occurred in developing countries. Increased life expectancy resulting from better medicines and healthcare account for the increase in world population. Unfortunately many people have died due to famine – lack of food. Developed countries have a more stable population.

The world population is expected to almost double by 2100. The **increasing human population requires increased food yield**. Therefore it is necessary to change and improve farming methods to produce food for the expanding population.

World population growth

After taking all of human history
for the population to reach one billion,
it took only a little over a century
to reach two billion in 1930.
The third billion was added in just
30 years, the fourth in only 15 years.

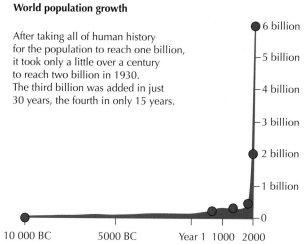

Fig 18.1 *The human population remained fairly stable until recently. The dramatic increase in growth is called a population explosion*

Make the link – Numeracy

In mathematics you may have learned how to interpret graphical trends.

Lines which are parallel to the x-axis indicate that the trend is occurring very slowly.

Make the link – Biology

In Area of study 3, Chapter 14, you learned about interspecific and intraspecific competition. These types of competition occur naturally in nature.

Pesticides reduce competition between crop plants and other organisms. This is an example of how human activities can reduce biodiversity.

Hint

Remember that due to the increasing human population farmers have to increase food yield. They do this by adding chemicals such as fertilisers and pesticides. Make sure you can explain why these chemicals increase food yield.

It is hoped that this could be achieved by using intensive farming practices which can involve the use of fertilisers and pesticides. Fertilisers are chemicals which add nutrients to the soil which are needed by plants. One **chemical** needed by plants is **nitrate**. Nitrates are used to increase **crop yield**.

Crop yield is increased when pesticides are used in farming. **Pesticides** are chemicals used to **kill animals** and **plants** which feed on crops. This is because less of the crop is eaten while it is growing meaning that more can be harvested for human use.

Intensive farming

Intensive farming is used to increase food production. Farmers grow crops known to produce a high yield. There is a heavy reliance on chemical fertilisers and pesticides.

Features of intensive farming methods

*   Competing plants are removed

    Soil is ploughed before planting. Unwanted plants that grow from naturally dispersed seeds are killed using pesticides. This allows more of the nutrients added to the soil, by applying fertilisers, to be taken up by the crop. However pesticides kill plants and so reduce biodiversity.

*   Herbivorous animals are removed

    Pesticides are sprayed onto the crop, killing animals that would eat the crop. The plants are less damaged and photosynthesise more. More energy can be transferred to the intended consumer (fig 18.2).

Fig 18.2 *Moth caterpillars eat the leaves of cherry trees and reduce the yield of cherries*

Fig 18.3 *Field sizes were small in the past. The land was used to grow the different crops required to feed both the people and the animals that lived on the farm*

Fig 18.4 *A huge area of land is devoted to growing the single plant type*

Make the link – Biology

In this chapter you will learn that GM crops were designed to increase the yield of food produced. There are only two GM crop plants authorised for cultivation in the EU. One is an insect-resistant maize plant and the other a potato with modified starch content. Neither of these types of plants is suitable for growing in Scotland.

Make the link – Social studies

The Industrial Revolution changed life for many people in Britain. People moved off the land and into cities. This resulted in the formation of transport links. What effects do transport links have on the environment?

Intensively farmed crops

Traditionally in Britain farming was done on a small scale. People grew a variety of crops. The crops they grew provided the food for them and their animals. Anything left over was sold. The problem was that there was never much left over (fig 18.3). As the population grew, farming methods changed. Farmers switched to intensive farming methods to provide food for the increasing population.

Monocultures are an example of intensive farming of plants. It is a type of farming where only one variety of crop is grown. The plants are normally genetically identical. Huge areas of land are used to grow the single crop (fig 18.4). Monocultures rely heavily on chemicals such as pesticides and fertilisers.

Biology in context

Many countries have a history of growing 'cash crops'. These are crops that are grown for their high monetary value. The problem is that they tend to be grown in countries where people do not have enough food to eat. Valuable land that could be used to feed the population is turned over to growing crops such as coffee and tea, which can be sold abroad for money.

GO! Activities

Activity 3.18.1 Working individually

1. State why the human population needs to increase its production of food.

2. Name two chemicals that can be used to increase food production.

3. For each of the chemicals named, explain why they are used in the production of food.

Activity 3.18.2 Working individually

The information contained in the passage below is adapted from an article on the Forestry Commission website.

Many people feel that monocultures reduce biodiversity. The Forestry Commission Scotland (FCS) puts forward an alternative point of view when discussing forest in Dunoon. The FCS admits that a type of spruce tree called Sitka when grown in monocultures 'can drain the soil of nutrients and crowd out other plants' when the tree is densely planted.

However, the FCS argues that when the spruce is planted as a monoculture, the trees form a dense canopy that prevents light reaching the forest floor. This provides a suitable habitat for many animals such as deer and foxes. The dense tree trunks provide them with cover and provide a sheltered habitat. The canopy also provides a habitat for many bird species such as goldcrests and siskins. These birds are the food source for birds of prey, which feed from the canopy on these smaller birds.

Gales coming from the Atlantic are strong enough to bring down some trees. The FCS states that when this happens, the trees are left uprooted to provide insects with a habitat. This is beneficial as small birds feed on the insects. When the land is harvested, and the trees are felled, many dormant seeds begin to grow due to the increase in soil temperature. Plants that grow when the spruce trees are felled include mosses, ferns and heather. Some of these seeds may have been dormant for up to 20 years in the leaf litter.

1. State a disadvantage mentioned in the passage of growing spruce trees in a monoculture.

2. Give two examples of mammals that obtain a sheltered habitat from spruce trees.

3. Name two birds that can be found living in the canopy.

4. Suggest why other plants are unable to grow on the forest floor.

5. Suggest why these forests may have a high biodiversity of birds of prey.

6. Suggest why birds such as goldcrests may move into the habitat shortly after a storm.

7. Name three plants that may lie dormant in the leaf litter.

8. Suggest which abiotic factor prevents these plants from germinating before the spruce trees are felled.

9. Predict the effect of felling the spruce trees on plant biodiversity.

I can:

- State that as the population of humans increases this requires an increased food production.

- State that to increase food production chemicals such as fertilisers and pesticides can be used in farming.

- State that fertilisers are used in farming as they contain the chemical nitrate.

- State that nitrates are used as they increase crop yield.

- State that competition between plants and crop plants can reduce crop yield.

- State that crop plants can be eaten by animals and so the food is not available for humans.

- State that pesticides can be used to kill plants and animals which are in competition with crop plants and so increase crop yield.

Chemical elements

All organisms are made from chemical elements. These elements include carbon, hydrogen, oxygen and nitrogen in the form of **nitrates**. Plants and animals need to take in these elements to stay alive. The elements are used to produce all the chemicals needed by the organism for growth. Nitrates can be found **dissolved in soil water**. Plants **absorb** nitrates through their roots. Nitrates then travel around the plant in xylem vessels. This allows the plant to make amino acids where they are needed by the plant.

The importance of nitrates to plants

Plants need nitrates to make **amino acids**. Amino acids are the building blocks for proteins. Therefore, the plant will carry out chemical reactions to change nitrates into amino acids. These amino acids are then joined together to make much larger protein molecules. Therefore, amino acids in plants are **synthesised** into **plant proteins**. Examples of proteins synthesised by the plant include enzymes, hormones and membrane proteins.

Plants lacking nitrates will suffer stunted growth. This will decrease the food yield of the plant. To **increase** the nitrate content of the soil farmers will add **fertilisers containing nitrates** to the soil. These nitrates dissolve in the soil water and can be absorbed by plant roots.

Some plants have special structures on their roots to help them use nitrogen gas from the air (fig 18.5). These plants can change nitrogen gas into nitrates in these root structures.

The importance of nitrates to animals

Animals are **consumers**. This means that they can obtain the nitrogen they need by consuming **plants** or **other animals**. This gives animals the source of amino acids they need to synthesise protein in their bodies. Animals must eat plants or other animals to obtain the amino acids they need for **protein synthesis**. Like plants, animals must make enzymes, hormones and membrane proteins.

Make the link – Biology

In Area of study 2, Chapter 11, you learned that plants have transport systems. The xylem vessels transport water, and chemicals which are dissolved in the water, from the roots to the leaves.

Make the link – Biology

In Area of study 1, Chapter 3, you learned how proteins are made in cells. Proteins are made from chains of amino acids. Nitrates are needed by plants to make amino acids.

Hint

Remember that proteins are made from chains of amino acids. These amino acids are joined together on the ribosomes.

Fig 18.5 *Root nodules on leguminous plants contain bacteria. The bacteria are able to convert nitrogen gas into nitrates, which can be used by the legume to produce amino acids*

🔍 Hint

Remember that nitrates are used to produce amino acids.

Amino acids are used in protein synthesis to make hormones, enzymes and membrane proteins.

🌳 Biology in context

To increase nitrates in the soil in a natural way, farmers may plant legumes. These are plants such as peas, beans or clover. These plants can convert nitrogen gas into nitrates. Organic farmers do this so they can avoid using chemical fertilisers. This means that they are able to grow a more valuable crop to sell as they are able to avoid the cost of buying fertilisers.

⁘ Make the link – Biology

In Area of study 1, Chapter 1, you learned the structure of the cell membrane. The cell membrane contains proteins which are produced from amino acids.

GO! Activities

Activity 3.18.3 Working individually

1. Name the substance that plants use to produce amino acids. Explain how plants obtain this substance.

2. Explain why plants require amino acids.

3. State the method animals use to obtain amino acids.

4. Explain why fertilisers are added to soil.

Activity 3.18.4 Working in pairs

An investigation was carried out using barley and pea plants.

Four pots each containing five seeds were planted. The contents of the pots are shown in the table.

(Continued)

⁘ Make the link – Health and Wellbeing

In food technology, you may have learned that carbohydrates, proteins and fats are three food groups that are needed for good health. Plants are the initial source of all proteins in food chains.

🔍 Hint

Remember that farmers add fertilisers to increase the nitrate content of the soil. This is important as it allows plants to produce higher yields.

Pot number	Contents of pot	
	Seed planted	Nutrient missing from soil
1	Barley	None
2	Barley	Nitrogen
3	Pea	None
4	Pea	nitrogen

The seeds were allowed to grow for 4 weeks. The heights of the seedlings were recorded.

Type of plant	Barley		Pea	
Nutrient missing from soil	None	Nitrogen	None	Nitrogen
Average height of seedling (cm)	30.0	24.0	34.0	30.6

(a) State the reason for including pots 1 and 3 in the investigation.

(b) Give two variables not mentioned that need to be controlled to make the experiment valid.

(c) Suggest how the reliability of the results could be improved.

(d) Calculate the percentage decrease in the average height of the barley and pea seedlings when grown without nitrogen.

(e) State which plant grew best when nitrogen was missing from the soil.

I can:

- State that plants absorb nitrates that are dissolved in soil water.

- State that the nitrate content of the soil can be increased by adding fertilisers.

- State that plants use nitrates to produce amino acids.

- State that the amino acids produced are synthesised (built up) into plant proteins.

- State that plants and animals are consumed by other animals to obtain amino acids which they then synthesise into proteins.

Fertilisers

Fertilisers are chemicals that are added to soil to improve plant growth. Fertilisers have different minerals added to improve the plant growth. A mineral added to the soil in fertilisers is **nitrates** (fig 18.6).

Fertilisers can **leach** from the soil into nearby **fresh water**. Leaching adds **extra nitrates** to the fresh water which are **unwanted**. Leaching occurs when rain washes excess fertiliser off the soil. The unwanted nitrates are used by tiny plants called algae which live in fresh water. These plants reproduce very quickly as they have a plentiful supply of nitrate. The **increase** in algal population can lead to **algal blooms** forming.

Fig 18.6 *Heavy machinery is used to spray fertilisers on farmland. This can lead to soil becoming compacted*

Algal blooms

Algae are tiny plants that live in water. They need nutrients to grow. When they grow in an uncontrolled manner they can form an algal bloom. Uncontrolled growth occurs in good growing conditions. The conditions needed for growth are a supply of minerals, light and warm temperatures (fig 18.7).

Algal blooms cover the surface of the water and **reduce light levels** reaching green plants under the surface. This prevents the plants gaining light. Plants are unable to carry out photosynthesis. The result of this is that **aquatic plants** are **killed**.

The dead aquatic plants become **food for bacteria** which live in the water. The bacteria also feed on **dead algae**. The plentiful supply of food allows the bacteria to dramatically **increase in numbers**. The high numbers of bacteria respire aerobically and use up **large quantities of oxygen**. This leads to a **reduction** in the availability of oxygen to **other organisms** which then also die. This reduces biodiversity in the fresh water.

Fig 18.7 *Algal blooms prevent light reaching green plants living under the water surface*

Reducing problems caused by using fertilisers

Due to the need to provide food for the increasing human population farmers need to use fertilisers. If farmers were to use natural fertilisers such as manure and compost, the risk of algal blooms would be reduced. This is because these natural fertilisers do not dissolve in water as easily and are therefore less likely to be washed into rivers.

🌳 Biology in context

Algal blooms can have an adverse effect on human health. Toxins produced by the algae can cause allergic skin reactions or contact dermatitis when swimming in contaminated water. Swallowing small volumes of this water can cause headaches, stomach cramps, nausea and diarrhoea.

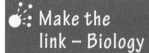

Make the link – Biology

In Area of study 3, Chapter 16, you learned that plants need water for photosynthesis. You also learned about factors which limit photosynthesis.

Plants living in water, aquatic plants, have a plentiful supply of water. This means that water is not a limiting factor.

This means that algae living in fresh water will reproduce rapidly if they receive a supply of nitrates. Algae use the nitrates to allow them to produce membrane proteins.

Make the link – Biology

In Area of study 2, Chapter 7, you learned about the process of mitosis. Mitosis is a type of cell division. It is used by multicellular organisms for growth and repair. Can you use your knowledge of mitosis to explain why algal cells are able to reproduce rapidly?

Make the link – Biology

In Area of study 1, Chapter 5, you learned that genetic engineering is a process that allows genes from one organism to be transferred into another organism. These techniques have been used to produce genetically modified rice plants. Tomatoes with longer shelf lives and blight-resistant potatoes have also been produced.

Farmers could use less fertilisers by switching back to more traditional farming methods. Traditional farming methods such as crop rotation used less chemical fertilisers.

In crop rotation, a different crop is grown each year. This is beneficial, as different crops have different mineral needs. Planting crops with deep roots one year and crops with shallow roots the next year gives the soil nutrient levels a chance to recover. This is a time consuming method. Therefore, modern farmers may use **genetically modified (GM) crop** plants.

Crop plants that have been genetically modified receive genes from other organisms. The genes they receive improve their ability to grow and produce a high food yield. Some crop plants have also been genetically modified to allow them to use nitrates more efficiently. This means that they produce a high yield using less fertilisers. It is hoped that future genetic modification of crop plants will be able to reduce the effect of intensive farming on the environment by **reducing the use of fertilisers**.

Genetically modified (GM) rice plants

Research has shown that some varieties of rice plants can be genetically modified to make more efficient use of nitrogen. This is an exciting development as it has been estimated that the human population will reach nine billion by the year 2050. Growing GM rice could ensure food security as rice yield could be increased even though fertiliser use was decreased.

This would be beneficial as the use of chemical fertilisers has been shown to have a negative effect on the environment. The negative effects include algal blooms, which can lead to greatly reduced oxygen levels in fresh water. Excess nitrogen can also be released into the atmosphere and contribute to harmful greenhouse gases.

GM rice has been engineered to contain a gene found naturally in barley. The gene that has been transferred into rice is known to aid the uptake of nitrates by barley roots. The gene increases the surface area of the roots. It is also thought to increase the quantity of amino acids in the roots. Amino acids are thought to have a role in increasing the size of the rice grains. Rice plants containing this gene are therefore better able to uptake nitrates than rice which has not been genetically modified. They will also produce a larger yield (fig 18.8).

Fig 18.8 *Genetically modified rice uses less fertiliser to obtain a larger food yield*

GO! Activities

Activity 3.18.5 Working individually

1. Explain how unwanted nitrates can end up in fresh water.

2. Name the type of plants, which live in fresh water, which use these extra nitrates.

3. Give the term that is used to describe what is formed when these plants reproduce rapidly.

4. Explain why the light available to underwater plants might decrease when extra nitrates are added to fresh water.

5. Explain why bacteria numbers may increase in these conditions.

6. State the substance, needed by other aquatic organisms, which is reduced by the bacteria.

7. Name the type of crops that can be grown to reduce the need for fertilisers in farming.

Activity 3.18.6 Working in groups

Six conical flasks were set up to investigate the effect fertilisers had on the growth of algae found in a pond.

A single sample of pond water was collected. The pond water was shaken and then each flask had the same volume of pond water added.

Each flask had a different volume of fertiliser added. The volumes added are shown in the table.

Make the link – Biology

In Area of study 1 Chapter 4 (pages **48–50**), you learned of the importance of amino acids in the formation of proteins. Remember that plants need to make proteins. They make protease enzymes to break down proteins into amino acids.

Make the link – Health and Wellbeing

Organic food is grown without using fertilisers.

Nutrients needed by plants are supplied by adding manure.

Why is organic food more expensive than intensively farmed food?

Hint

Remember that dead algae and plants provide food for bacteria. Due to the increased food available to the bacteria they dramatically increase in numbers. Then large numbers of bacteria use up oxygen. This is what reduces biodiversity.

(Continued)

Flask	Volume of fertiliser added (cm³)	Dry weight of algae produced (g)
1	0	0.05
2	0.5	0.10
3	1.0	0.18
4	1.5	0.20
5	2.0	0.34
6	4.0	0.34

1. State the variable that was investigated in this experiment.
2. Give two variables that would have to have been kept constant to make this a valid experiment.
3. Suggest how the reliability of the experiment could have been improved.
4. Explain why conical flasks were used instead of beakers.
5. Suggest why the pond water was shaken before being added to the flasks.
6. Suggest why the flasks were left on a warm, sunny window ledge.
7. Explain why one flask had no fertiliser added.
8. Collect a piece of graph paper. Draw a line graph to show the relationship between the volume of fertiliser added and the dry weight of algae produced.
9. Describe the relationship between the volume of fertiliser added and the dry weight of algae.
10. Explain why the dry weight of algae may have levelled off at volumes above 2 cm³.
11. State how many times higher the dry weight of algae produced at 1.5 cm³ of fertiliser added was than the control.
12. Suggest why the dry weight of algae was used to compare the results rather than the fresh mass.

Activity 3.18.7 Working in pairs

Scientists are trying to reduce the effect of fertilisers on the environment by genetically modifying crop plants. They hope to produce crops which can produce a high food yield while using as little fertiliser as possible.

Your task is to produce a newspaper article where you explain recent developments to the readers. Your article should be between 200 and 300 words long. You should use the internet or other resources available to you to find out more information.

You should include:

1. The name of the crop plant which has been genetically modified.
2. The source of the gene that has been transferred into the crop plant.
3. The reason why the crop plant has been modified to contain the gene.
4. The advantages to the environment of genetically modifying the crop plant.
5. Reasons why some people may be against genetically modifying the crop plant in this way.

I can:

- State that fertilisers added to crops can leach into fresh water.
- State that this adds extra and unwanted nitrates into the fresh water.
- State that algae are plants that live in fresh water.
- State that the extra nitrates cause the algal population to increase which can cause an algal bloom.
- Explain that an algal bloom can reduce light levels which reach underwater plants, and this kills the plants living in the water.
- Explain that bacteria live in fresh water and the dead plants and dead algae provide food for the algae allowing them to rapidly increase in number.
- Explain that the extra food allows the bacteria to greatly increase in number.
- Explain that this is a problem as the bacteria use up large quantities of oxygen and greatly reduce the oxygen available for other aquatic organisms.
- State that crops which have been genetically modified (GM) can be used to reduce the need for fertilisers.

Pesticides

Pesticides are chemicals that are **sprayed** onto **crops** to reduce competition from other organisms (fig 18.9). **Crop yield** is increased when **pesticides** are used in farming. Pesticides are chemicals used to **kill animals** and **plants** which feed on crops. This is because less of the crop is eaten while it is growing

Hint

Remember that 'cide' means 'killer of'. Therefore, pesticides are killers of pests. Pests are any organisms that reduce crop yield.

Fig 18.9 *Spraying pesticides reduces the effects of competition. They can have a negative effect on human health and biodiversity*

> **Make the link – Biology**
>
> In Area of study 3, Chapter 14, you learned that competition occurs between organisms. The function of pesticides is to reduce competition.

> 🔍 **Hint**
>
> Remember that a toxic substance is one that has an adverse effect. A toxic substance causes harm to, or kills, organisms.

> 🔍 **Hint**
>
> Remember that lethal means that the substance is able to cause the death of organisms.

> **Make the link – Health and Wellbeing**
>
> Organic food is grown without the use of pesticides. Instead, pests are removed from the crop plant by hand. Can you explain why organic food is more expensive than intensively farmed food?

meaning that more can be harvested for human use. There are different types of pesticides. Two examples are given below.

1. Herbicides are used to kill plants which are in competition for resources with the crop plant.

2. Insecticides are used to kill insects feeding on crop plants. This makes more of the crop available to humans. It reduces competition between insects and humans.

If competition between the crop plant and other organisms is reduced then the crop yield is increased. However, pesticide use has been responsible for a decrease in biodiversity. This is because some pesticides sprayed onto crops **bioaccumulate in the bodies of organisms over time**. This was observed after the use of the insecticide DDT.

Pesticides and the food chain

DDT is a powerful insecticide first used in farming after World War 2. From 1972 it was banned for agricultural use in the United States. It has since been banned worldwide.

The ban was necessary as the chemicals in DDT were found to be non-biodegradable. This means that the chemicals in DDT could not be broken down by the animals that ate them. Therefore every time food contaminated with DDT was eaten, DDT built up in the animal. This is known as **bioaccumulation**. Bioaccumulation is the **build-up** of **toxic substances** in living **organisms**.

Bioaccumulation of DDT occurred along food chains. Organisms at each link in the food chain were found to accumulate DDT in their tissues.

DDT entered the food chain in plants. Consumers of plants took DDT into their bodies. Plants tolerated the DDT due to the low levels of the chemical. However, consumers of plants were found to have more DDT in their bodies than was present in the plants. The level had accumulated. The level in the bodies of consumers continued to accumulate through to the final consumers in the food chain or web. The final consumers contained the highest level of DDT (fig 18.10).

DDT was responsible for a large reduction in predatory bird numbers. This was not only due to the increase in toxicity. A large concentration of DDT was found to make it difficult for birds to absorb calcium. It was also thought to alter hormones needed for reproduction. Birds laid eggs with thin shells, and the chicks hatched before they were strong enough to survive.

As certain pesticides pass along **food chains, toxicity increases and can reach lethal levels**.

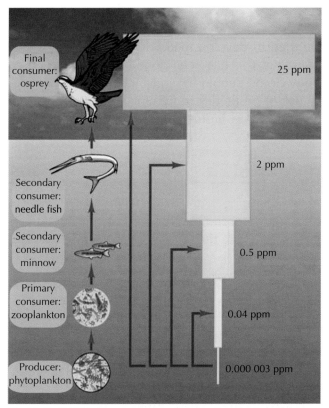

Fig 18.10 *DDT is bioaccumulated in the tissues of organisms as it passes along the food chain. The concentration of DDT in the sea water is expressed as parts per million (ppm)*

Final consumer: osprey — 25 ppm

Secondary consumer: needle fish — 2 ppm

Secondary consumer: minnow — 0.5 ppm

Primary consumer: zooplankton — 0.04 ppm

Producer: phytoplankton — 0.000 003 ppm

🍄 Biology in context

Ospreys are birds that became extinct in Scotland early in the twentieth century (fig 18.11). They re-colonised naturally in 1954. By the 1970s there were still only 14 breeding pairs. It is thought that the slower than expected increase could have been due to bioaccumulation of DDT and other pollutants affecting the birds' breeding success. The Osprey Centre at Loch Garten is a conservation site.

Make the link – Biology

Many soldiers serving in the Gulf War have been treated for Gulf War Syndrome. Symptoms described by British soldiers include headaches, memory loss and muscle or joint pain. There is no firm explanation for the cause of Gulf War Syndrome but many people believe that it could have been caused by pesticides sprayed on tents to kill mosquitoes. Since the soldiers slept in tents sprayed in this way they may have taken lots of pesticides into their bodies.

Make the link – Biology

In Area of study 3, Chapter 17, you learned that the final consumers in food chains and webs are shown at the top of pyramids of numbers. These final consumers are most affected by the build-up of toxic pesticides as they eat a lot of prey organisms which contain the pesticide.

🔍 Hint

Accumulate means to gradually build up or acquire. Organisms furthest up the food chain accumulate the highest levels of pesticides in their tissues.

Fig 18.11 *DDT bioaccumulated in the tissue of ospreys. This led to thinning of eggshells*

Alternatives to the use of pesticides

Scientists have tried to develop methods to reduce the use of pesticides in farming. This has been done as pesticides can be expensive to buy. Some people have also raised concerns over the largescale use of pesticides in farming. Alternatives to the use of pesticides include **biological control** and **genetically modified (GM) crops**.

Biological control

The use of intensive farming methods has been shown to have a harmful effect on the environment. **Biological control and GM crops may be alternatives to the use of fertilisers and pesticides** in farming.

Biological control is seen as an alternative to using pesticides. It relies on natural predation instead of chemicals. A natural predator that feeds on the pest is released. The number of pests is reduced. An added bonus is that the cost of this control method is relatively cheap after the initial setup costs. Another benefit of using predators to combat pests is that it is a natural method and can be used in organic farming. Predators will remain in the area as long as there is food available.

Need for biological control

Many farmers use chemical pesticides to control pests. There are many disadvantages to this method.

- Chemical pesticides are non-specific. This means that beneficial insects may be killed. These insects might act as pollinators.

- Pest species may become resistant to the pesticide. All the pests in the area will eventually survive as the resistance is passed on to offspring.

- Chemical pesticides may enter the food chain. Bioaccumulation may occur and biodiversity may be reduced.

- Spraying pesticides is a health risk to those living near agricultural land.

Advantages of biological control

- The predator that is introduced is specific to the pest. This means that biodiversity may not be reduced.

- After the initial setup costs it may be much cheaper to control pests using biological control.

- There are no problems of chemicals contaminating the environment or the food chain.

- Most pests do not become resistant to the treatment, as no chemicals are used. The same method of control can be used year after year.

Disadvantages of biological control

- Its use in small confined areas, e.g. greenhouses, has found more success than in fields. It is too easy for the predators to leave a field environment.

- It can initially be very expensive to practise the technique due to the high development costs.

- Some introduced predators have become pests themselves. This happened with the cane toad (fig 18.12).

- Not all the pests will be killed. Some remain and still cause damage. Even with biological control, pesticides may still need to be used.

Biological control in action

An early example of biological control occurred when the virus *Myxoma* was released into rabbit populations. The virus caused myxomatosis. Rabbits infected with the virus die a week after infection. The rabbits develop sores to their eyes and become blind.

The virus killed many rabbits, so crop damage was reduced. An unexpected result was that rabbits became immune to the virus and it became an unsuccessful method of control.

Ladybirds eat a variety of insects (fig 18.13). They have been introduced to kill aphids and scale insects. Aphids feed on the sap of plants. They can introduce pathogens into plants. Plants infested with aphids tend to produce few, small flowers. Crop plants need to produce flowers to produce the useful product. It is thought that one ladybird is able to eat up to 50 greenfly in a day and 5000 over its life.

Genetically modified (GM) crops

Genetically modified crops are crop plants that have had their genetic material altered. This is done by inserting a gene from another organism into their genetic material. This is usually

Fig 18.12 *Cane toads were introduced in Australia to control cane beetles. They have become a pest species themselves*

Fig 18.13 *Ladybirds are the natural predators of aphids*

⁛ Make the link – Biology

You studied food chains and food webs in Area of study 3 Chapter 14 (pages **226–227**). Remember if a predator is a successful hunter, the numbers of prey decline. The prey is the pest species. For this reason biological control should not become a problem so long as the predator is carefully chosen.

⁛ Make the link – Biology

In Area of study 1, Chapter 5, you learned about the processes involved in genetic engineering. These techniques are used to insert the Bt toxin gene into crop plants to reduce the need to use pesticides.

Biology in context

Biological control can also be used to control weeds. A weed is a plant that grows out of control. This is the case with Japanese knotweed. In 2010 a trial was undertaken where an insect native to Japan was released into selected sites in England. It was hoped that the insect would feed on the sap of the weed and stunt its growth. If it is successful it could save the UK economy more than £150 million a year.

 Make the link – Religious and moral education

You may have studied ethics. Ethics involves deciding the rights and wrongs of situations. Can you apply your knowledge of ethics to breeding organisms to control the growth of other organisms?

 Hint

GM crops and biological control are used as alternatives to pesticides. These technologies cost less and are potentially less damaging to the environment and human health.

done in the laboratory. Once the gene has been successfully transferred, the crop plant will produce seeds which also contain the new trait. These seeds can be used by farmers. The aim of genetic modification is to alter the crop plant so that it will produce a higher food yield.

Advantages of GM crops

Scientists have developed GM crops for use in farming because:

- Some crop plants now have an inbuilt resistance to insect pests. If the pest eats the crop plant it will die. This is beneficial as it reduces the need to spray the crop plant with chemical pesticides.

- Some crop plants now have an inbuilt resistance to the chemical weed killers used in farming. This means that less weed killer needs to be sprayed onto the crop.

- The crops contain genes that are not normally found within their species. It is possible to transfer genes from bacteria, animals and other plants into crop species.

- The crop plant will produce a higher yield than the non-genetically modified crop plant.

- It results in less exposure to pesticides for farm workers and nonharmful insect species.

Disadvantages of GM crops

Some people question the use of GM crops because:

- They fear it could lead to a loss of biodiversity due to the reduction in the number of weed species. This is a concern as weed species provide animals with a food source.

- They fear that through natural selection, weeds that do survive being sprayed by weed killer will be impossible to get rid of in the future.

Bt toxin

Bacillus thuringiensis is a bacterium that produces proteins that are toxic to many insect species. There are thought to be over 200 different types of toxic protein produced by the different strains of the bacterium. These toxic proteins kill a wide range of different types of insect. Several species of crop plant have been modified to contain the Bt toxin gene. These include tomatoes, potatoes, corn and soya (fig 18.14).

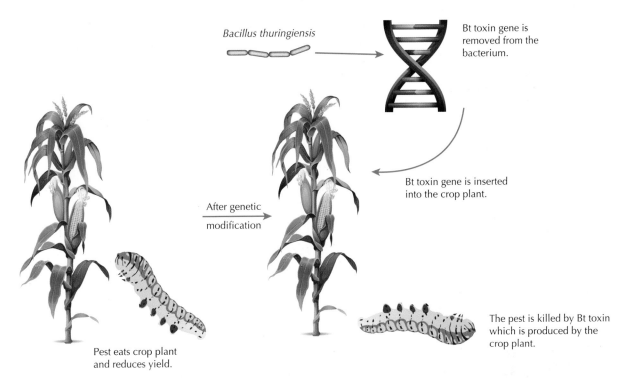

Bacillus thuringiensis

Bt toxin gene is removed from the bacterium.

Bt toxin gene is inserted into the crop plant.

After genetic modification

The pest is killed by Bt toxin which is produced by the crop plant.

Pest eats crop plant and reduces yield.

Fig 18.14 *GM crops are engineered to contain a gene that kills insect pests*

ᴳᴼ! Activities

Activity 3.18.8 Working individually

1. Explain why pesticides are sprayed onto crops.
2. Name two different types of pesticides used in farming and state why each one is used.
3. Give the term used to describe the build-up of pesticides in organisms.
4. Suggest why DDT accumulated in the organisms at the top of the food chain.
5. Explain the effect of DDT on reproduction in birds.

Activity 3.18.9 Working individually

The table below shows the average levels of DDT in human body fat of people living in the United States during the period 1942–1982.

Year	DDT level in human body fat (mg/g fat)	Year	DDT level in human body fat (mg/g fat)
1942	0	1970	11.6
1950	5.3	1972	9.2
1956	10.6	1974	6.7
1962	12.6	1976	5.5
1963	10.3	1980	0
1968	12.5	1982	0

(Continued)

1. Suggest a reason for the lack of DDT in human body fat in 1942.
2. State the number of years it took for the DDT level in human body fat to double after it was first detected.
3. Suggest a reason for the lack of DDT from 1980 onwards.
4. DDT is an insecticide. Suggest a reason why it might have been found in human body fat.

Activity 3.18.10 Working individually

The following table shows the results of an investigation into DDT and bioaccumulation in a food chain.

Organism	DDT concentration (ppm)
Phytoplankton	0.000003
Zooplankton	0.04
Minnow	0.5
Needle fish	2
Osprey	25

1. Draw a food chain for the organisms present.
2. State the percentage increase in the DDT concentration between the minnow and the osprey.
3. State the ratio of DDT in zooplankton compared to the osprey.
4. Give the number of times more DDT is present in the tissue of needle fish than in zooplankton.

Activity 3.18.11 Working in pairs

The caterpillar moth (*Cactoblastis*) has been introduced to kill cacti (*Opuntia*).

1. Look up the following web address:

   http://www.northwestweeds.com.au/sample-page/cactoblastis-biocontrol/
2. Produce a newspaper article of around 200–300 words informing readers of the success in using *Cactoblastis* in biological control. Your article should be one A4 page in size.

   Include the following information:

   • The natural habitat of the moth.
   • The plant that it has been used to control. Explain why the use of the moth has been effective.
   • A photograph of both the moth and the plant.
   • A description of the moth that would allow it to be recognised.
   • A description of the life cycle of the moth including information on the following stages – eggsticks, larvae and moths.
   • Natural predators of the moth.

Activity 3.18.12 Working in pairs

Many crop species have been genetically modified to contain the Bt toxin gene. Your task is to research the use of the Bt toxin gene in tomato plants. You should present your findings as a newspaper article to support the use of Bt toxin in farming. You should include:

- An explanation of where the Bt toxin gene is naturally found.
- An explanation of how the Bt toxin gene is transferred to crop plants.
- A list of crop plants that have been successfully genetically modified to include the Bt toxin gene.
- An explanation of why tomatoes have the Bt toxin gene.
- The benefits to the environment of growing tomatoes that contain the Bt toxin gene.

I can:

- State that pesticides sprayed onto crops can build up in the bodies of other organisms over time.

- State that pesticides can contain chemicals that are toxic to other organisms.

- State that this build-up of toxic substances is known as bioaccumulation.

- Explain that these pesticides increase in toxicity as they are passed along the food chain and can reach lethal levels.

- State that biological control uses natural predators to control pests in farming.

- State that genetically modified (GM) crops and biological control can be used as alternatives to pesticides in the production of crop plants.

19 Evolution of species

Mutation

> **Hint**
>
> Remember that spontaneous means that mutations occur without any apparent cause.

A **mutation is a random change to the genetic material** of an organism. Mutations are rare, random and **spontaneous**. They cause permanent change to the DNA. The result of a mutation is that the genetic message is changed. This can alter the phenotype of the organism.

Some mutations are more significant than others. Mutations that occur in sex cells are inherited by offspring. Mutations **are the only source of new alleles**. Mutations that occur in body cells can lead to cancer.

Causes of mutations

Mutations occur naturally and spontaneously. They are rare and occur infrequently. A mutation changes the organism's DNA. Naturally occurring mutations occur when a cell divides. Prior to cell division, DNA replicates, or copies, itself. If the newly formed copy is not identical to the original DNA, a mutation has occurred.

Certain **environmental factors** are called **mutagenic agents**. A mutagenic agent is something that **can increase the rate** at which mutations occur. **Radiation** and **some chemicals** are environmental factors which are known to increase the rate of mutation.

Being exposed to mutagenic agents can increase the rate of mutations. Examples of environmental factors that are mutagenic agents are:

- **Radiation** – UV light, X-rays and gamma rays.
- **Some chemicals** – mustard gas, colchicine, caffeine, formaldehyde and chemicals in cigarettes.

Research scientists may expose cells artificially to mutagenic agents. They do this to artificially increase the rate of mutation. It does not cause a specific mutation. By exposing cells to mutagenic agents, scientists hope to find cures for diseases.

Types of mutations

Mutations can be divided into three groups depending on the effect they have on the survival of the organism that the mutation occurs in.

Mutations conferring an advantage

A mutation can result in an **advantage to survival** for the organism. These are called **advantageous** mutations. They result in an improvement in the organism's phenotype. These mutations are very rare but also very important. They are the source of all new alleles. Advantageous mutations may lead to the evolution of new species as they allow natural selection to occur.

An advantageous mutation can be seen in the peppered moth *Biston betularia*. A random gene mutation produced a melanic moth (fig 19.1). This black moth was camouflaged from predators in industrial areas where trees were no longer covered with lichens and had been blackened with soot. It therefore survived

Make the link – Biology

Mutations could occur when new cells are being produced. You studied the process of producing new cells in Area of study 2, Chapter 7 (pages **92–105**). DNA codes for the proteins required by the cell. You studied the role DNA plays in the production of proteins in Area of study 1, Chapter 3 (pages **40-47**).

Make the link – Social studies

'Agent Orange' is a powerful herbicide sprayed by Americans on vegetation during the Vietnam War. It was hoped that the enemy would be more visible when the trees dropped their leaves. It was then discovered to be a powerful mutagenic agent. No one realised the birth defects that it would cause. Can you think of any other mutagenic agents that have been used during warfare?

Hint

Gametes are sex cells. If a mutation occurs during gamete production new alleles may be formed. Alleles are the variations that exist for one particular gene. This new allele may be expressed in the organism's phenotype. The phenotype is the physical appearance of the organism. Read more about phenotypes in Area of study 2, Chapter 10.

Fig 19.1 *Peppered moth*

Fig 19.2 *Cystic fibrosis is an example of a disadvantageous mutation. Sufferers have difficulty breathing due to the thick, sticky mucus they produce*

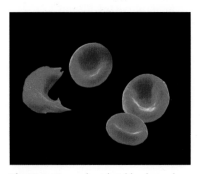

Fig 19.3 *Normal and sickle-shaped red blood cells*

to breed and so passed on its advantageous allele to its offspring. It is considered a classic example of natural selection in action.

The failure of all spindle fibres during cell division in plants can give rise to an advantageous mutation. The resulting plant is called a polyploid. They are known to possess increased resistance to disease. They are larger in size and grow more vigorously than their non-polyploid relatives. Many cereal plants are polyploid.

Mutations conferring a disadvantage

A mutation can result which is a **disadvantage to survival** for the organism. These are called **disadvantageous** mutations. If the mutation occurs in genes that code for vital enzymes the organism may not survive. Some mutations can lead to sections of a chromosome being lost. This could mean that the organism lacks vital genes. Such a mutation would therefore be a lethal mutation. Disadvantageous mutations may lead to extinction.

Cystic fibrosis is an example of a condition caused by a disadvantageous mutation. A mutation in the gene that controls the movement of salt and water in and out of the cells within the body causes cystic fibrosis. The condition affects the lungs and digestive system, which become clogged with thick mucus. This results in difficulty breathing and digesting food (fig 19.2).

Sickle-cell anaemia is another example of a condition caused by a disadvantageous mutation. People with one mutated allele are normally healthy. They may even have a selective advantage over people with two normal alleles in areas where malaria is present. People with two sickle cell alleles have the disease and the disadvantage. During exercise, their red blood cells take on a sickle shape and can be trapped in small blood vessels. This results in clots which cause painful blockages. The sickle cells are unable to carry as much oxygen as normal cells (fig 19.3).

Neutral mutations

Mutations with no effect on the survival of the organism are called **neutral** mutations. These mutations may occur in part of the DNA that does not code for a protein. They can also occur in a gene that is not expressed in the phenotype. An incorrect amino acid can be incorporated into the protein but it does not alter the function of the protein. An extra toe would be a neutral mutation since an extra toe does not alter the organism's ability to survive in its environment (fig 19.4). More than one gene is involved in the formation of the toe.

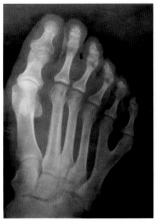

In Area of study 2, chapter 10, you learned how genetic information is inherited by offspring from their parents. Read over this chapter to make sure you understand how genetic conditions are inherited.

Ψ Biology in context

During pregnancy, the foetus can be tested for the presence of mutations. The test is called an amniocentesis. A sample of cells is removed from the amniotic fluid and the chromosomes present are analysed.

Fig 19.4 *This person has an extra toe. There is neither an advantage nor a disadvantage to this. It is a neutral mutation*

GO! Activities

Activity 3.19.1 Working individually

1. Give a definition for the term mutation.
2. State three ways that mutations can influence the survival of organisms.
3. Explain what the term spontaneous means with reference to mutations.
4. Explain why mutations are important in evolution.
5. Give two examples of environmental factors that can increase the rate of mutation.
6. State the term used to describe factors which can increase the rate of mutation.

Activity 3.19.2 Working individually

Mutations can be caused by being exposed to radiation and from chemicals. A study was carried out into the sources of radiation and how much each source contributed to the total exposure.

It was found that natural radioactivity in the air accounted for 38% of the total radiation. The ground and buildings exposed people to exactly half of the natural radioactivity in the air. Surprisingly, 16% of the radiation came from food and drink. Medical X-rays accounted for 13%. Cosmic rays from space accounted for 1% more than medical X-rays.

1. Present the information given above as a table.
2. State the percentage contribution that comes from natural radioactivity and medical X-rays.

Activity 3.19.3 Working in pairs

Cigarette smoke is known to contain chemicals that are mutagenic agents and so increase the rate of mutations in smokers.

Produce an information leaflet that could be given to smokers to explain the risk of developing cancer as a result of smoking.

(Continued)

You should:

- State the number of mutations needed for a normal cell to turn into a cancer cell.
- Explain why most types of cancer are more common in older people.
- Explain why smoking could lead to cancer.
- State the number of cancers that are caused by smoking.
- Explain why heavy smokers are more likely to develop cancer.
- Include a photograph of healthy lungs, and lungs that have been damaged by smoking.

Your leaflet should be one folded A4 page.

I can:

- State that a mutation is a random change to the genetic material of an organism.
- State that mutations occur spontaneously.
- State that mutations are the only source of new alleles.
- State that the rate that mutations occur can be increased by certain environmental factors.
- Give radiation and some chemicals as environmental factors that can increase the rate of mutations.
- State that some mutations are neutral, which means that they do not affect the organism's survival.
- State that mutations which give the organism a better chance of survival are said to confer an advantage to survival.
- State that mutations which give the organism a poorer chance of survival are said to confer a disadvantage to survival.

Variation

Make the link – Biology

In Area of study 2, Chapter 10, you learned how variation arises in species. A variation is a difference that exists between members of the same species.

These variations occur due to new alleles which are inherited by offspring. Mutations are the only source of new alleles.

All members of a population show variation. A variation is simply the differences between members of the population. Variation results from differences in genetic information the organisms contain. These differences occur because of **new alleles produced by mutations**. Some of the new alleles may result in **animals and plants being better adapted to their environment**. Adapted animals and plants have inherited characteristics from their parents that make them well suited to survive in their niche.

In ideal conditions, all members of the population could survive until old age. However, most organisms fail to survive to reproductive age. **Variation within a population** allows the population to **evolve**. Evolution occurs over many generations and usually takes a long period of time. Evolution occurs in response to **changing environmental conditions**. The members of the population which survive the changing environmental conditions survive to reproductive age. This means they may be able to pass on the favourable new alleles to their offspring.

Changes to environmental conditions could include natural disasters such as drought, floods and fires. As a result, the habitat changes. These changes to the habitat could alter the population in several ways.

Animals – Over time the quantity of food, water and availability of shelter may change. Animals also must survive changes to diseases and predators in their environment.

Plants – Over time the quantity of soil nutrients, light and water may change in the environment. Plants also must survive changes to diseases and herbivores in the environment.

Organisms that have adaptations allowing them to survive will pass on these adaptations to their offspring.

Adaptations

Organisms that are well adapted to their niche cope better with changes that occur in their environment. They are better at competing for food, escaping from predators and fending off disease. They are more likely to survive and produce offspring. The favourable alleles that allowed them to survive will be passed on to the offspring.

Adaptations to extreme habitats

Special adaptations are needed for survival in extreme habitats. Deserts are hot and dry, and there may be very little rain. There is very little shade, and daytime temperatures are very hot. At night it is very cold.

The desert rat (fig 19.5) is a nocturnal animal. This means that it is active at night when the temperature is lower. It does not sweat and produces little urine.

Camels (fig 19.6) have large feet to prevent them sinking into the sand. They also have a double layer of eyelashes to reduce damage to their eyes.

Cacti are good examples of plants that are adapted to live in extreme habitats (fig 19.7). Their stems store water. Instead of leaves they have spines to reduce water loss by transpiration.

Hint

Remember that mutations are the only source of new alleles. This means that beneficial mutations must occur to allow organisms to adapt to their environment.

Make the link – Biology

You learned how alleles are passed on to offspring in Area of study 2, Chapter 10, pages **136–157**. You should be able to work out how even a recessive allele gradually changes the population if it is advantageous.

Fig 19.5 *Desert rats are nocturnal mammals*

Fig 19.6 *Camels are well adapted to living in hot and dry conditions*

Make the link – Social studies

You may have studied deserts in detail. You will know that the conditions are harsh and that water is not readily available. What other requirement needed by plants may be missing from desert soil?

Fig 19.7 *Desert plants are well adapted to living in arid conditions*

Hint

Transpiration is the evaporation of water from leaf surfaces. Water is lost through the stomata. Read more about transpiration in Area of study 2, Chapter 11.

Biology in context

Humans are not well adapted to living in desert conditions. Each year people die in Death Valley National Park in the US. There have been recent examples of satellite navigation taking people on an incorrect route. They run out of water and are not adapted to living in the harsh conditions.

Activities

Activity 3.19.4 Working individually

1. State the importance of new alleles, produced by mutation, to animals and plants.
2. State the term that is used to describe the differences that exist between members of the same population. Explain why this is important.
3. State the meaning of the term adaptation.
4. 'Populations evolve over time' – explain what this statement means.
5. Explain why it is necessary for populations to evolve over time.

Activity 3.19.5 Working individually

The diameter of a barrel cactus was measured before and after rainfall (fig 19.8). The measurement was found to be 32 cm before the rain fell and 44 cm after rainfall.

1. State the percentage increase in the diameter of the cactus after rainfall.
2. Suggest an adaptation that this type of cactus has to allow it to increase in diameter.
3. The leaves of these cacti are reduced to spines to reduce water loss. Suggest another advantage of having spines instead of leaves.

Fig 19.8 *A clump of barrel cacti*

Activity 3.19.6 Working in pairs

The photograph below shows a moth pollinating a flower at night (fig 19.9).
Think about the adaptations that the flower shows.

1. The flower is white. Explain how this adaptation helps the moth to act as the pollinator.

2. The flower usually produces a strong and sweet-smelling scent. Explain how this adaptation helps the moth to act as the pollinator.

3. The moth has a long nectar tube. Explain how this adaptation helps the moth to act as the pollinator.

Fig 19.9 *Moths can transfer pollen from flower to flower*

Activity 3.19.7 Working in pairs

Galapagos finches were observed by Charles Darwin. He thought that a few finches had been blown onto the Galapagos Islands from the mainland. We now think that mutations occurred in these ancestral finches that were beneficial. This meant that the ancestral species was able to evolve over time and exploit different ecological niches.

Study the diagram on the next page and answer the questions below.

1. Give the diet of the ancestral finch.

2. State the minimum number of mutations that must have occurred to allow the modern finches to change their diets.

3. Explain how these mutations allowed the finches to change their diet.

4. Give the habitat of the ancestral finch.

5. Give three other habitats that the modern finches live in.

6. Suggest a reason for modern finches changing their habitat.

(Continued)

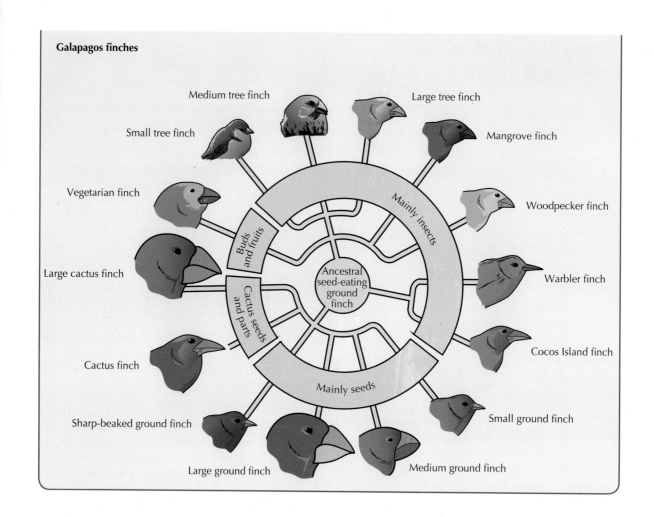

Galapagos finches

Medium tree finch

Small tree finch

Large tree finch

Mangrove finch

Vegetarian finch

Woodpecker finch

Large cactus finch

Warbler finch

Cactus finch

Cocos Island finch

Sharp-beaked ground finch

Small ground finch

Large ground finch

Medium ground finch

Mainly insects

Buds and fruits

Cactus seeds and parts

Ancestral seed-eating ground finch

Mainly seeds

I can:

- State that all members of a population show variation.

- State that new alleles, produced by mutations which confer an advantage to survival, are responsible for the variation in populations.

- Explain that these new alleles can result in animals and plants becoming better adapted to their environment.

- Explain that an adaptation is a characteristic inherited by the organism which makes it well suited to its niche and so increases its ability to survive in its niche.

- Explain that variations present within a population allow the population to evolve over time in response to changes to the population's environmental conditions.

Natural selection

The idea of natural selection was first put forward by Charles Darwin and Alfred Wallace in 1858. Darwin published his book 'On the Origin of Species' where he explained his concept of natural selection (fig 19.10).

Natural selection is based on the fact that all organisms reproduce to produce offspring to carry on their species. **Species produce more offspring than the environment can sustain.** This leads to a 'weeding out' process. Natural selection is often referred to as **survival of the fittest**. This is because natural selection occurs when there are **selection pressures**. Selection pressures affect an organism's ability to survive. They include factors such as disease, predation and food shortage. Natural selection results in only the best-adapted organisms surviving to reproduce. Selection pressures are different for different individuals of the species. Those not adapted may die. The **best-adapted individuals in a population survive to reproduce**. Reproduction allows them to **pass on their favourable alleles** to their offspring. Favourable alleles are ones that **confer the selective advantage**. They are the alleles that allowed the parents to survive in their environment. The offspring inherit the favourable alleles and are more likely to survive to reproductive age themselves. This means they are also able to pass on their favourable alleles. By this means, the **frequency of the favourable alleles** in the population **increases**.

Darwin travelled the world on the HMS Beagle. He studied organisms and noted patterns that he observed. His studies allowed him to conclude:

1. Species produce many more offspring than the environment can sustain.

2. All offspring show variations. Some variations may be advantageous, and others may be disadvantageous.

3. This leads to a struggle for survival. For example they struggle to obtain food and mates, to avoid predation and disease. These are the **selection pressures**.

4. Only the best-adapted organisms with the favourable alleles are able to survive to reproductive age.

5. These organisms pass on the alleles that confer the selective advantage.

6. The offspring produced experience selection pressures.

7. Only those that are best adapted survive to reproductive age.

8. Gradually the whole population changes to possess the alleles that confer the selective advantage (fig 19.11).

Fig 19.10 *Charles Darwin proposed the theory of evolution*

Make the link – Biology

In Area of study 3, Chapter 14, you learned that organisms are in competition with each other. The competition is most intense between members of the same species – intraspecific competition.

This is the type of competition which illustrates the term 'survival of the fittest' – only those organisms best adapted to the environment survive to reproduce.

Hint

Remember that all organisms produce more offspring than the environment can sustain. Sustain means support. So, there will always be too many organisms born that can be supported by the environment.

Make the link – Social subjects

In geography, you may have learned about famine and the causes of famine. Famine occurs as the environment is unable to support the population.

Hint

Remember that natural selection is still continuing in organisms. New variations occur due to mutation. If the mutation has a selective advantage, the whole population will gradually evolve to possess the mutant allele. This takes a very long time to happen.

Any breeding pair of animals will produce more offspring than the environment can sustain.

The offspring show variation due to inherited characteristics.

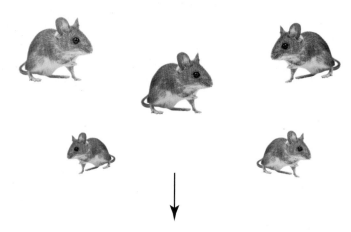

The offspring experience selection pressures.
Only the best-adapted individuals in a population survive.

Competition for resources, disease, food availability, grazing and predation are examples of selection pressures which lead to death for the least well-adapted organisms.

The best-adapted individuals are able to reproduce.
They pass on the favourable alleles that confer the selective advantage to their offspring.

The favourable alleles increase in frequency within the population.

Fig 19.11 *Natural selection, or survival of the fittest, occurs when there are selection pressures and results in only the best-adapted organisms surviving to reproduce*

Through these stages, Darwin was able to explain the evolution of organisms best adapted to their environment. It is important to remember that characteristics of a population change gradually from generation to generation. Only the fittest individuals survive to pass on the favourable alleles. The process takes time.

Natural selection in action

People may believe that once an organism has adapted to its environment there is no need to change further. This is not the case. The environment is constantly changing so organisms have to constantly evolve. Natural selection continues to change the characteristics of a population. It does this over a long period of time. There are some examples where the process of natural selection takes place fast enough to be observed.

Peppered moth

A classic example is that of the peppered moth *Biston betularia*. Two forms of the moth exist. The light form is dominant, and the dark one arose from a mutation (fig 19.12). The light one possessed a selective advantage, as it was able to camouflage against lichen-covered tree trunks (fig 19.13) where it rested during daylight. The dark moth had a selective disadvantage, as it was easily spotted by predators.

During the Industrial Revolution in 19th-century Britain burning coal produced gases which killed lichens, and surfaces became encrusted with soot. The dark moth was now camouflaged. It had the selective advantage. It survived to reproduce and pass on the dark allele (fig 19.12). Gradually the population changed in industrial areas so that more dark moths appeared.

The 'Clean Air Act 1956' reduced pollution released by factories. The lichens on trees in industrial areas recovered. Gradually the number of dark moths decreased as they were easily seen by predators and eaten.

Resistant bacteria

Bacterial infections have been commonly treated with antibiotics since the 1950s. This has led to the development of resistant strains of bacteria. Natural variations in the populations of bacteria meant that some were not killed by the antibiotic. The sensitive bacteria were killed, leaving the antibiotic-resistant bacteria to thrive. They had a selective advantage. New types of antibiotics were produced, but over time they have all become less effective.

Fig 19.12 *Two forms of the peppered moth exist. Dark moths possess a selective advantage on trees where lichens have died*

Fig 19.13 *Light moths possess a selective advantage on lichen-covered trees*

🌳 Biology in context

MRSA is a type of bacterial infection. It is often carried in the nostrils and throat, and on the skin. It is difficult to treat. Through natural selection it has become resistant to many widely used antibiotics. People in hospital are most at risk. This could be because their immune systems are weaker. Also, after an operation there may be an open wound that could become infected. Good hygiene has reduced the number of cases in recent years.

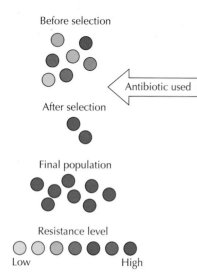

Before selection

After selection

Final population

Resistance level

Low   High

Fig 19.14 *Natural variation means that some bacteria have resistance before the antibiotic is used. After the antibiotic is used, all bacteria are resistant*

Rapid natural selection

MRSA

Bacteria such as MRSA have become resistant to antibiotics due to natural selection. Each bacterium shows variation. Some will have natural resistance to antibiotics. The antibiotic kills the sensitive bacteria. The remaining bacteria have a selective advantage and survive to pass on the resistance to their offspring. If these bacteria are exposed to the same antibiotic again they survive and multiply. Eventually, a population of bacteria evolves that is completely unaffected by the antibiotic. Since some bacteria can reproduce every 20 minutes, the whole population can become resistant in a very short period of time (fig 19.14).

GO! Activities

Activity 3.19.8 Working individually

1. Explain why organisms produce offspring.
2. Comment on the number of offspring an organism will produce with reference to its environment.
3. State a second term that can be used to describe natural selection.
4. Explain what is meant by the term selection pressures.
5. State the organisms in a population which survive to reproduce.
6. Explain why it is beneficial that these organisms survive to reproduce.
7. State what happens to the frequency of favourable alleles in a population as a result of these organisms surviving to reproduce.

Activity 3.19.9 Working in groups

Darwin's finches are found on the Galapagos Islands. They are thought to have evolved from one type of ancestral finch blown over from the mainland (fig 19.15).

1. State two quantitative measurements that could be taken of the birds' beaks. A quantitative measurement is one that can be measured.
2. From the diagram, give one environmental factor that could have led to this variation.
3. Suggest which finch is most similar to the ancestral finch. Explain your choice.
4. Suggest which finch is most dissimilar to the ancestral finch. Explain your choice.

Make the link – Biology

In Area of study 2, Chapter 10 (page **143**), you learned about dominant and recessive alleles. Mutations are the source of new alleles. Could you explain why a recessive allele may not show up in the phenotype of an affected individual?

Make the link – Religious and moral education

You have learned the theory of natural selection. This is not the view taken by many religions. Some people believe that life on Earth exists exactly as God created it. Could you put forward an argument to support or refute the theory of natural selection?

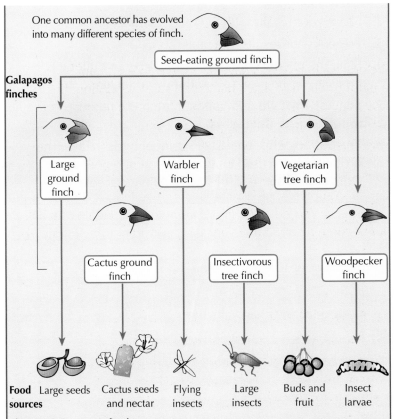

Fig 19.15 *Darwin's finches*

Activity 3.19.10 Working in groups

To do this activity you will need:

- Bulldog clips of different sizes
- A Petri dish for each group member
- A small food container filled with broth mix
- A stopclock.

(a) Each member of the group should select a bulldog clip of a different size.

(b) Each member has a Petri dish on the desk beside them.

(c) Place the food container filled with broth mix in the middle of the group.

(d) Each group member should use the bulldog clip to remove the broth mix from the dish.

(e) You are only allowed to remove one piece of the mix at a time.

(f) You have to place the piece of the mix into the Petri dish.

(g) The time allowed is 30 seconds.

(h) Your aim is to collect as many pieces of the mix as possible.

1. After 30 seconds, count the pieces of broth mix of each type you have.

(Continued)

2. Share your results with the other members of your group.

3. Present your results as a table.

4. State the relationship between the size of bulldog clip and size of the piece of broth mix collected.

5. Suggest what would happen if the broth mix was replaced with dried peas.

6. The bulldog clips are used to model birds' beaks. The broth has been used to model different types of food. Explain how these results apply to natural selection.

I can:

- State that all organisms produce offspring to carry on their species.

- State that species produce more offspring than can be sustained by the environment.

- Explain that natural selection is based on the idea of a weeding out process where only the best-adapted members of the species survive to reproduce. This is known as survival of the fittest.

- Explain that natural selection occurs when there are selection pressures that only some members of the species are adapted to – these are the members of the species which survive.

- Explain that the individuals in the population which are best adapted to the environment pass on their favourable alleles to their offspring when they reproduce.

- State that the favourable alleles are the alleles which confer a selective advantage.

- Explain that natural selection results in an increase in the frequency of the favourable alleles in the population.

Species

Organisms that are able to interbreed and produce fertile offspring are called a **species**. Being fertile means that the offspring are able to reproduce and produce offspring themselves.

Some closely related organisms are able to interbreed, but the offspring are sterile. Horses and donkeys are closely related. They are able to interbreed and produce offspring but the offspring are infertile and cannot reproduce. A cross between a male horse and a female donkey produces a mule (fig 19.16).

Fig 19.16 *A mule is sterile and unable to produce offspring*

Lions and tigers are also closely related and produce offspring called ligers. The offspring are infertile (fig 19.17).

Organisms that are not closely related are unable to produce offspring with each other.

Process of speciation

Speciation is a process that allows new species to develop. **Part of the population becomes isolated by an isolation barrier**. There are three types of isolation barrier: **behavioural, geographical and ecological**. Isolation means that members of the same species follow different evolutionary paths due to natural selection. The organisms with the genes best suited to the selection pressures survive to breed and so pass on their favourable alleles.

Speciation requires several processes to take place (fig 19.18).

Fig 19.17 *A liger is unable to produce offspring, as it is sterile*

🔍 Hint

Remember that for speciation to occur the following processes are needed: Isolation followed by Mutation followed by Natural selection.

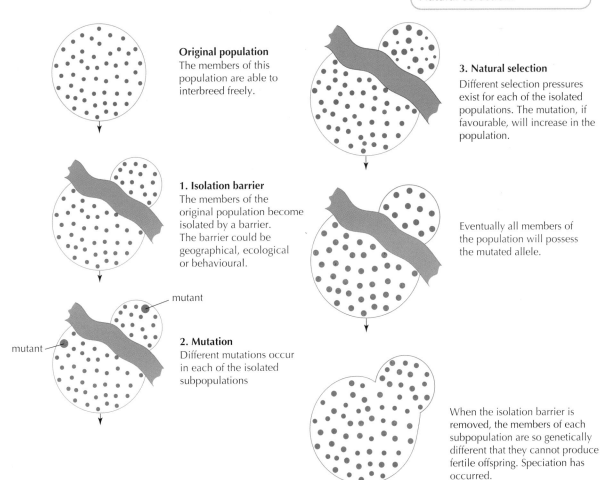

Original population
The members of this population are able to interbreed freely.

1. Isolation barrier
The members of the original population become isolated by a barrier. The barrier could be geographical, ecological or behavioural.

mutant

mutant

2. Mutation
Different mutations occur in each of the isolated subpopulations

3. Natural selection
Different selection pressures exist for each of the isolated populations. The mutation, if favourable, will increase in the population.

Eventually all members of the population will possess the mutated allele.

When the isolation barrier is removed, the members of each subpopulation are so genetically different that they cannot produce fertile offspring. Speciation has occurred.

Fig 19.18 *The processes involved in speciation*

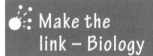
1. **Isolation**

   Part of the original population becomes isolated. This prevents the members of the new subpopulations from interbreeding with each other and the original population. The exchange of genes by interbreeding is prevented. The isolation can be brought about by:

   - **Geographical** barriers – these could be seas, rivers, mountains or deserts, which prevent subpopulations from meeting and breeding.

   - **Behavioural** barriers – such as subpopulations of animals with mating displays which don't attract those in other subpopulations or subpopulations of flowering plants which have a different flowering period from other subpopulations.

   - **Ecological** barriers – in which different subpopulations have different habitat preferences such as moisture level or salinity level, so don't meet and breed with other subpopulations.

2. **Mutation**

   In each of the isolated subpopulations **different mutations** occur. Remember mutations are spontaneous, rare and random. Therefore, the mutations that occur on either side of the isolating barrier will be different. Mutations create new, and different, alleles in each of the subpopulations. These alleles did not exist in the original population.

3. **Natural selection**

   Each subpopulation will be exposed to **different selection pressures**. These selection pressures could be due to differences in climate, predation or diseases. The organisms which possess the new, mutated, favourable alleles will possess a selective advantage. **Natural selection** therefore selects **different mutations** in each subpopulation.

   If the mutation gives members of the subpopulation a selective advantage it will be selected for. Those organisms not possessing the advantageous mutation will be selected against. Only the new favourable mutation will be passed on to offspring.

4. **Time**

   Evolution continues over many generations. Mutations and natural selection will drastically change the gene pool of each population. They will become two distinct groups. If the barrier is removed, they will no longer be able to interbreed and produce fertile offspring. Therefore, each **subpopulation will evolve until** they become so genetically different **they are two different species.**

GO! Activities

Activity 3.19.11 Working individually

1. Explain what is meant by the term species.

2. Give the term that is used to describe the process that allows a new species to develop.

3. Name three different types of isolation barrier.

4. Explain why isolation is needed in the formation of a new species.

5. Explain why mutation is needed in the formation of a new species.

6. Make a comment on the mutations that occur in each of the isolated subpopulations during the process of speciation.

7. Explain the role of natural selection in the formation of new species.

8. Explain how it is possible to tell that two, different, species have formed from the original species.

Activity 3.19.12 Working in pairs

Isolation is important in the process of speciation. Members of the same species share a common gene pool. The frequency of the alleles of genes is maintained by random mating. During the process of speciation, members of the same species are prevented from interbreeding by isolation barriers. The diagram below shows the land mass on Earth (fig 19.19).

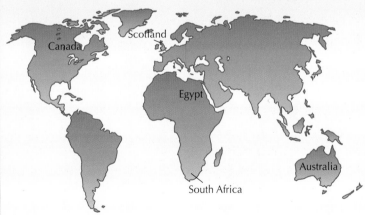

Fig 19.19 *Land mass on Earth*

Isolation barriers can be geographical barriers, behavioural barriers or ecological barriers. Geographical barriers include seas, rivers, mountains and deserts. Ecological barriers result from variations in temperature, water availability and pH of different ecosystems. Behavioural

Y Biology in context

On the island of Arran, two of Scotland's rarest tree species are found. The trees do not exist anywhere else in the world and there are only thought to be a couple of hundred of them. These trees are the Arran whitebeam (first observed in 1897) and the Arran cut-leaved whitebeam (first observed in 1952). The Arran cut-leaved whitebeam is thought to have arisen from a cross between the Arran whitebeam and a rowan tree. They have arisen due to isolation leading to speciation. As their habitat is reduced so the number of these rare trees has declined. WWF classes these trees as dangerously near to extinction.

Make the link – Biology

Speciation has led to insect resistance to toxins produced by GM crops. You have already learned about genetic engineering (Area of study 1 Chapter 5, pages **64–67**) and should know how the insect toxin was introduced into the crop plant.

Make the link – Social studies

Archaeologists look for clues to events that happened in the past by carrying out digs to look for ancient relics. How do biologists gain evidence to support the theory of evolution?

(Continued)

barriers include sex cells that cannot fuse, mating displays that don't attract members of another population or flowering at different times of the year.

1. Give three types of isolation barriers.

2. Suggest the type of barrier that could prevent a plant living in Canada interbreeding with a plant in Scotland. Explain your choice.

3. Suggest the type of barrier that could prevent a plant living in South Africa interbreeding with a plant living in Egypt. Explain your choice.

4. Suggest the type of barrier that could prevent a plant living in northern Australia interbreeding with a plant in southern Australia. Explain your choice.

Activity 3.19.13 Working in pairs

Go to www.kilda.org.uk and follow the link to the 'Wee St Kilda Guide'. From there, find out about the St Kilda wren.

St Kilda is an island found in the North Atlantic Ocean. It is over 100 miles away from the Scottish mainland. The St Kilda wren is found there. Research the ways in which the St Kilda wren is different from the mainland wren.

Include:

- Differences in overall size, wing length, bill thickness and leg thickness.
- Appearance, including colouring and markings.
- Size and weight of eggs.
- The natural habitat of the birds.

1. Suggest reasons for these differences.

2. Predict what might happen if the St Kilda wren was introduced to the Scottish mainland.

I can:

- State that a species is a group of organisms that can interbreed to produce fertile offspring.

- State that speciation is the name given to the process where two new species are formed from one original species.

- State that speciation occurs when members of a population become isolated from each other by an isolation barrier to form two subpopulations.

- Give geographical barriers, ecological barriers and behavioural barriers as examples of isolation barriers.

- List oceans, rivers, seas and mountains as examples of geographical barriers.

- List pH, salinity and different habitats as examples of ecological barriers.

- List mating behaviour, courtship rituals and gamete incompatibility as examples of behavioural barriers.

- Explain that after isolation, different mutations occur in each subpopulation.

- Explain that selection pressures for each subpopulation will be different meaning that natural selection will select different mutations as favourable in each subpopulation.

- Explain that because of natural selection, each subpopulation evolves differently.

- Explain that since each subpopulation evolves differently, they will become so genetically different that they become two different species.

Life on Earth – Review questions

Section A

1. Competition between organisms is most intense when it is

   A interspecific competition between members of the same species

   B intraspecific competition between members of the same species

   C interspecific competition between members of different species

   D intraspecific competition between members of different species

2. Two examples of biotic factors are

   A grazing and wind speed

   B grazing and predation

   C predation and wind speed

   D predation and humidity

3. Indicator species indicate levels of pollution by their presence or absence. Mayfly nymphs indicate unpolluted water. The table below shows the results of a survey carried out on four rivers.

   Mayfly nymphs would be found in river ……… .

River	Oxygen level	Number of bacteria in water
A	Low	Low
B	High	Low
C	High	High
D	Low	High

4. The graph shows the numbers of foxes and rabbits found in a forest over a 10-week period. Rabbits are the prey of foxes.

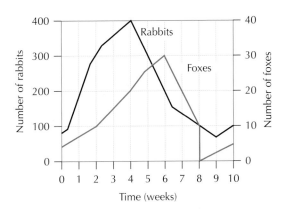

The ratio of foxes to rabbits in week 2 is

A 1:3

B 30:1

C 1:30

D 10:300

5. The raw materials for photosynthesis are

   A carbon dioxide and water

   B oxygen and water

   C carbon dioxide and sugar

   D oxygen and sugar

6. The graph below shows the effect of light intensity, concentration of carbon dioxide and temperature on the rate of photosynthesis.

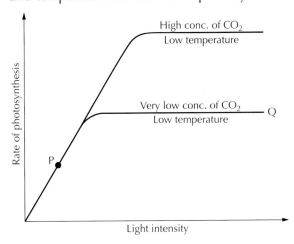

Which factors are limiting the rate of photosynthesis at points P and Q on the graph?

	P	Q
A	Light	Carbon dioxide
B	Light	Temperature
C	Carbon dioxide	Temperature
D	Temperature	Light

7. Between each level in a food chain, energy is lost as

   A heat only

   B movement only

   C heat and movement

   D heat, movement and undigested materials

8. Information about fertiliser usage in Scotland is shown in the graph below.

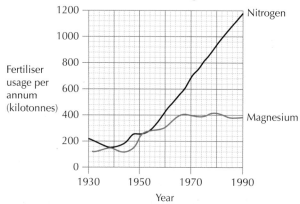

   Which of the following statements about magnesium usage is correct?

   A Magnesium usage per year reached 1200 kilotonnes.

   B Magnesium usage per year never rose above 420 kilotonnes.

   C Maximum magnesium usage per year was 440 kilotonnes.

   D Magnesium usage never exceeded nitrogen usage.

9. Mutations are random changes to an organism's genetic material.

   Mutations are

   A advantageous only

   B disadvantageous only

   C neutral only

   D the only source of new alleles

10. Speciation describes the evolution of new species. A new species is considered to have evolved when a population

    A   has members that can no longer produce fertile offspring

    B   can no longer produce fertile offspring when interbreeding with the original population

    C   shows increased variation due to mutations

    D   shows decreased variation due to lack of mutations

Section B

1. The diagram below shows part of a woodland food web.

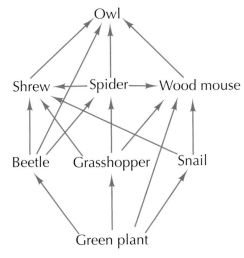

    (a) Use five organisms from the food web to construct a food chain below.

           _____ → _____ → _____ → _____ →

           _____   1

    (b) Decide if each of the following statements is **True** or **False**, and tick (√) the appropriate box.

        If the statement is **False**, write the correct word in the **correction box** to replace the word(s) underlined in the statement.

Statement	True	False	Correction
In this food web, grasshoppers are <u>carnivores</u>.			
In this food web, the wood mouse passes its energy to <u>spiders</u>.			

                                                                          2

(c) Explain why a decrease in shrew numbers may lead to an increase in wood mouse numbers.

_____

_____ 1

(d) Underline **one** option in each set of brackets to make the following sentence correct. 1

Beetles are the (predators/prey) of spiders. Woodmice are (herbivores/omnivores).

(e) Give the term used to describe the role a organism plays within a community.

_____ 1

2. The light intensity inside and outside a woodland was measured.

The results are shown in the table below.

| Month | Average daily light intensity (units) | |
	Outside woodland	Inside woodland
January	15	14
February	17	16
March	22	19
April	26	23
May	29	26
June	33	18
July	36	10
August	31	10
September	27	10
October	22	10
November	18	16
December	15	14

(a) Explain why the light intensities outside and inside the woodland showed the greatest difference between June and October.

_____

_____ 1

(b) State **one** further factor that could have been measured to compare the effect of abiotic factors on the distribution of organisms within and outside the woodland.

Give a possible source of error when measuring the abiotic factor you have **named** and explain how the effect of the error could be minimised.

Abiotic factor _____ 1

Source of error _____ 1

Method of minimising error _____

_____ 1

(c) Quadrats and pitfall traps could be used to sample organisms living in the woodland.

For either quadrats **or** pitfall traps give a possible source of error when using the technique and explain how the effect of the error could be minimised.

Source of error _____1

Method of minimising error _____

_____1

3. An investigation was carried out into the effect of competition on the survival of grass seedlings. Five identical Petri dishes were set up as shown in the diagram. Each dish contained a different number of grass seeds as shown in the table below.

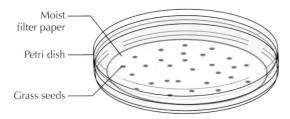

Each dish was left to grow for 7 days. The results of the investigation are shown in the table below.

Dish	Number of seeds sown	Number of surviving seedlings after 7 days	Percentage of surviving seedlings after 7 days (%)
1	10	10	100
2	20	18	90
3	40	32	80
4	80	48	60
5	100	35	35

(a) State the variable that was altered in the investigation.

_____ 1

(b) State **two** variables, not already mentioned, that need to be kept the same in this investigation.

1. _____

2. _____ 2

(c) Explain why the percentage of surviving seedlings after 7 days was calculated.

_____

_____ 1

(d) Suggest one abiotic factor that the grass seedlings may be competing for.

_____ 1

(e) Draw a **line graph** to show the number of seeds planted against the percentage of seedlings surviving after 7 days.

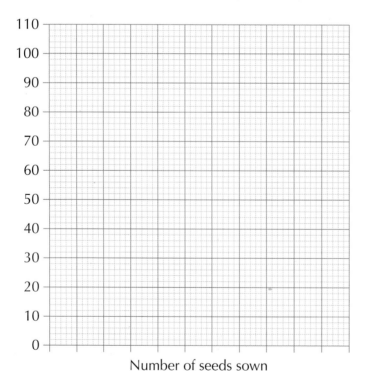

Number of seeds sown

2

4. An experiment was set up to measure the effect of light intensity on the rate of photosynthesis in the aquatic plant, *Elodea*.

The light intensity was varied using a dimmer switch on the bulb.

The rate of photosynthesis was measured by counting the number of bubbles released per minute.

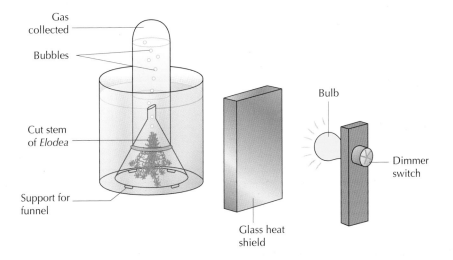

The results of the experiment are shown in the table below.

Light intensity (units)	Rate of photosynthesis (number of bubbles per minute)
2	4
4	12
6	28
8	48
10	48
12	48

(a) Describe the relationship that exists between the light intensity and the rate of photosynthesis.

_____

_____ 2

(b) Name two other factors which can also affect the rate of photosynthesis.

1. _____

2. _____ **2**

(c) State one product of the light reactions needed for the carbon fixation stage of photosynthesis.

_____ **1**

5. Fertilisers and pesticides are chemicals that are used in intensive farming of crop plants. For either fertilisers **or** pesticides, state why the chemical is used **and** describe the effect it has on the **environment**. Also, give a **suggestion** to reduce this effect on the environment.

_____

_____

_____

_____

_____

_____ **3**

6. A mutation is a random change to genetic material. Mutations can be caused by environmental factors such as some chemicals.

(a) Give **one other** environmental factor that can cause mutations.

_____ **1**

(b) Other than mutations, name **two** other processes that occur during the process of speciation.

1. _____

2. _____ **2**

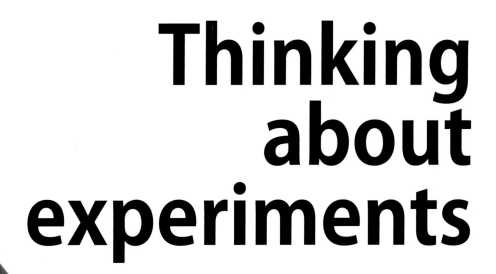

Thinking about experiments

1. Measuring enzyme activity

Cell biology – background and additional knowledge:

- Enzymes function as biological catalysts and are made by all living cells.

- Enzymes speed up cellular reactions and are unchanged in the process.

- The shape of the active site of an enzyme molecule is complementary to its specific substrate(s).

- Enzyme action results in product(s).

- Enzymes can be involved in degradation and synthesis reactions.

- Enzymes have specific substrate(s) and produce specific product(s).

- Each enzyme works best in its optimum conditions.

- Enzymes and other proteins can be affected by temperature and pH.

- Lactose is a milk sugar.

- Lactase is the enzyme that breaks down lactose into sugar molecules.

- Lactase can be **immobilised** (stopped from moving) by attaching it onto jelly beads meaning it can be easily separated from the product.

- The presence of sugars produced can be tested for by the use of clinistix.

Putting it in context:

- The diagram below shows how the enzyme lactase is used in the production of lactose-free milk.

- Lactose-free milk is made commercially for people who suffer from lactose intolerance so it has to be cost effective to produce.

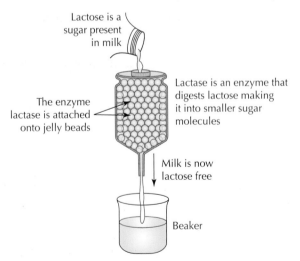

The aim of the experiment was to find out if temperature affects the activity of the enzyme lactase.

Results:

Temperature of milk and lactose (°C)	Time taken for sugar to be present (s)
10	72
20	50
30	46
40	25
50	78

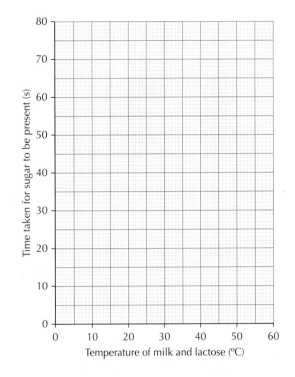

Collect a piece of graph paper and copy the axes on the right.

Use the information in the table above to plot a **line graph** of the results.

Design features of the experiment:

1. **Give** a reason why the enzyme is immobilised on the jelly bead.
2. **Predict** the time taken for sugar to be present if the experiment had been carried out at 1°C. **Explain** your prediction.
3. **State** the control for this experiment.
4. **Give** two variables, not already mentioned, that would have to be kept the same to make sure that the experiment was valid.
5. **Explain** how the milk could be kept at the temperature required for each experiment.
6. **Explain** how the results could be made more reliable.
7. **Predict** the time taken for sugar to be present if this experiment had been repeated at 60°C.

Analysis of results:

1. **State** how much longer it took for sugar to be produced at 20°C compared to 40°C.
2. **State** how many times more time was taken for sugar to be produced at 20°C compared to 40°C.
3. **State** the percentage decrease in time taken for sugar to be present at 40°C compared to 20°C.
4. **Calculate** the average time for sugar to be present.
5. **State** the relationship between the temperature and the time taken for sugar to be present.
6. **Explain** why the results should be drawn as a line graph.
7. **Give** a conclusion for this experiment.

2. Using a respirometer

Cell biology – background and additional knowledge:

- Respirometers can be used to measure rate of respiration.

- A respirometer is a device used to measure the rate of respiration of a living organism by measuring its rate of exchange of oxygen for carbon dioxide.

- A rate is a measure of the way a factor changes **in a given time period**.

  E.g. the increase in temperature in a room in **30 minutes**.

Putting it in context:

- The rate of respiration in germinating (growing) seeds can be measured using a respirometer as shown in the diagram below. The respirometers can be exposed to different temperatures to find out how temperature affects the rate of respiration in seeds which are germinating.

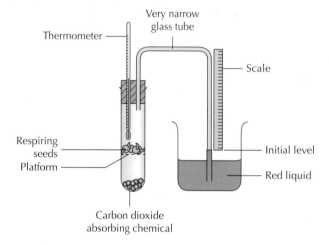

- Four identical respirometers were set up as shown above.

- Two respirometers contained respiring seeds. The other two contained dead seeds. The dead seeds had been boiled in water to denature their enzymes.

- Two respirometers were exposed to a temperature of 10°C. The other two were exposed to 30°C.

- The level of the red liquid was measured on the scale at the start of the investigation and after 30 minutes.

- The rise in the liquid level was due to the uptake of oxygen by the respiring seeds.

- The aim of the experiment was to investigate if temperature affects the rate of respiration of seeds.

- The results are shown in the table.

Results:

Respirometer	Temperature (°C)	Contents of tube	Rise in liquid level (mm)	Rate of oxygen uptake (mm per minute)
A	10	Respiring seeds	12	0.4
B	10	Dead seeds	0	0
C	30	Respiring seeds	36	1.2
D	30	Dead seeds	3	

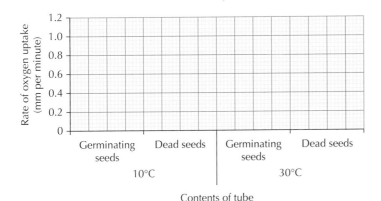

Collect a piece of graph paper and copy the axes on the left.

Complete the table for tube D. Use the information in the table above to draw a **bar chart** of the results.

Design features of the experiment:

1. **Give** the reason for adding red liquid to the beaker.

2. **Give** a reason for the presence of the scale found next to the glass tube.

3. **Explain** why the level of the red liquid had to be measured at the start and the end of the experiment.

4. **Name** a piece of apparatus that is not shown in the diagram which would be needed to measure the rate of respiration.

5. **Explain** why the respiring seeds are on a platform inside each tube.

6. **Explain** why a chemical to absorb carbon dioxide needs to be included in the tube containing the respiring seeds.

7. **Explain** why the glass tube connecting the red liquid to the boiling tube had to be very narrow.

8. **Give** the name used to describe tubes B and D. **Explain** why they were included in the experiment.

Analysis of results:

1. **State** how many times more the rate of oxygen uptake (mm per minute) was at 30°C compared to 10°C, for respiring seeds.

2. **State** how much more oxygen was consumed at 30°C compared to 10°C, for respiring seeds.

3. **State** the percentage increase in oxygen consumed at 10°C compared to 30°C, for respiring seeds.

4. **Calculate** the average rise in liquid level.

5. **State** the relationship between the temperature and the rate of oxygen uptake by the respiring seeds.

6. **Explain** why the results should be drawn as a bar graph.

7. **Give** a conclusion for this experiment.

3. Measuring transpiration using a potometer

Multicellular organisms – background and additional knowledge:

- Transpiration is the loss of water through leaves by evaporation.

- During transpiration water is lost by evaporation through the stomata, whose opening and closing is controlled by guard cells which are mainly found in the lower leaf epidermis.

- A potometer is a device used for measuring the rate of transpiration of a leafy plant shoot.

- There are two types of potometer – bubble and weight.

- The rate of transpiration is affected by several factors, including:

Factor	Effect on transpiration rate (when factor is increased)
Temperature	Increased
Humidity	Decreased
Wind speed	Increased
Surface area of leafy shoot	Increased

Putting it in context:

- The rate of transpiration can be measured using the apparatus shown below – a **bubble** potometer.

- An air bubble is created in the red water.

- As the plant transpires, red coloured water is drawn along the glass tube.

- The distance that the air bubble moves is measured, over a known period of time, to allow the rate of transpiration to be calculated.

Bubble potometer

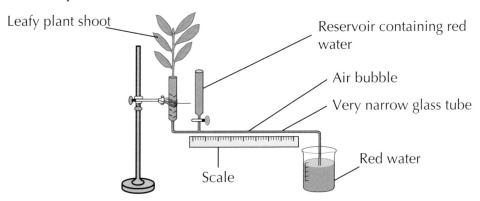

Leafy plant shoot

Reservoir containing red water

Air bubble

Very narrow glass tube

Scale

Red water

- The apparatus above is called a **bubble potometer**.

- The red water stored in the reservoir can be used to return the air bubble to the start of the scale so that water transpired by the plant in different conditions can be measured.

- The aim of this experiment was to investigate the effect of temperature on the rate of transpiration.

Results:

Temperature (°C)	Distance moved by the air bubble (mm)
5	12
20	24
30	48
40	72

This experiment was set up and then **12 hours** later the results were obtained.

Calculate the rate of water loss at each temperature.

5°C   20°C

30°C   40°C

Copy and complete the table below

Temperature (°C)	Rate of transpiration (cm³ of water per cm³ of leaf surface area per hour)
5	
20	
30	
40	

Design features of the experiment:

1. **State** a piece of apparatus, not shown on the diagram, which would be needed to measure the rate of transpiration.

2. **Give** two variables not already mentioned that would need to be controlled to obtain valid results.

3. **Explain** why a very narrow glass tube was used in this experiment.

4. **Suggest** why red dye was added to the water (this isn't normally done).

5. **Explain** the need for the reservoir of red coloured water.

6. **Explain** why Vaseline might be placed around the stopper where it meets the tube.

7. **Explain** why the units used to show the rate of transpiration are cm³ of water per cm³ of leaf surface area per hour.

Analysis of results:

1. **State** the temperature where the fastest rate of transpiration occurred.

2. **State** the temperature where the slowest rate of transpiration occurred.

3. **Predict** the effect on the rate of transpiration if the experiment had been repeated at 55°C.

4. **Calculate** the average rate of transpiration.

5. **Describe** the trend shown by the results.

6. **Give** a conclusion for the experiment.

Weight potometer

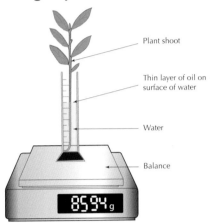

Plant shoot

Thin layer of oil on
surface of water

Water

Balance

8594 g

The apparatus above is called a **weight potometer** and is used to measure transpiration in leafy plant shoots or in plants. The balance is used to measure the weight of the potometer.

Design features of the experiment:

1. **Give** two measurements that would have to be taken to use this apparatus to measure the rate of transpiration.

2. **Explain** why it could be difficult to measure the effect of temperature on transpiration using this apparatus.

3. **Name** two factors that affect the rate of transpiration that could be investigated using the above apparatus.

4. **Explain** why a layer of oil has been placed on the surface of the water in the tube containing the leafy plant shoot.

4. Measuring abiotic factors

Life on Earth – background and additional knowledge:

- Abiotic factors include light intensity, soil moisture, pH and temperature.

- There are possible sources of error when measuring abiotic factors.

- There are techniques that can be used to minimise sources of error when measuring abiotic factors.

- Light intensity and soil moisture are both measured using a light moisture meter.

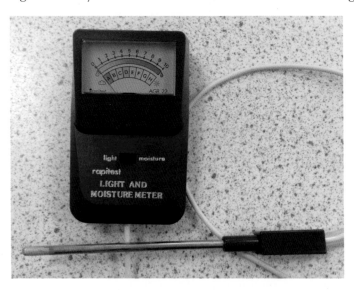

A light moisture meter is used to measure both light intensity and soil moisture. The switch must be set to the abiotic factor that you wish to measure

- pH is measured with a pH probe.

- Temperature is measured using a thermometer.

Putting it in context:

- An investigation was carried out to determine if soil moisture levels were different at the top of a slope compared to the bottom of the slope.

- A **transect line** was laid out from the top of the slope to the bottom of the slope.

- Quadrats were placed at 1 m intervals along the **transect line**.

- In each quadrat, the soil moisture reading was taken three times.

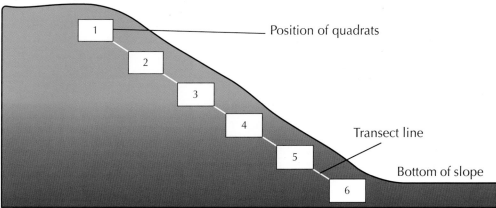

Results:

Quadrat	Soil moisture (%)			Average soil moisture (%)
	S1	S2	S3	
1	10	7	13	
2	15	17	13	
3	40	35	45	
4	63	5	22	
5	72	80	88	
6	90	80	82	

Design features of the experiment:

1. **Explain** why a transect line was used to place the quadrats.
2. **Explain** why the soil moisture readings were only taken inside each quadrat.
3. **Give** a source of error that could occur when using a soil moisture probe.
4. **Give** the technique that could be used to minimise the source of error that you have identified.
5. **List** three other abiotic measurements that could have been taken in this investigation.
6. For each of the abiotic factors you have identified, **list** an error that could be made when sampling.
7. For each of the errors you have identified, **give** a technique that could be used to minimise the effect of the error on the results.
8. **Explain** why six quadrats were used instead of one at the top and another at the bottom of the slope.

Analysis of results:

1. **Calculate** the average soil moisture content found in each quadrat.

2. **State** the trend shown by the average soil moisture content.

3. **Predict** the average soil moisture content which would have been found if a seventh quadrat had been placed after quadrat 6.

4. **Give** a conclusion for this investigation.

5. **State** the quadrat number where a sampling error may have occurred. **Explain** why you thought this was the sample site where a sampling error had occurred.

6. **Give** an aim for this experiment.

5. Measuring the distribution of a species

Life on Earth – background and additional knowledge:

- Biotic factors include predation, grazing, food availability, disease and competition for resources.

- Biodiversity and the distribution of a species can be influenced by biotic factors.

- When an agricultural crop is harvested the land is ploughed and left to the spring before planting begins again.

- Farmers wishing to grow organic crops must leave fields baron (abandoned) to allow time for the chemicals used in intensive farming to reduce to an acceptable level in the soil.

- Black grass is a weed which is causing problems to farmers.

- Black grass plants are not highly competitive, but their seed production is very high meaning populations can increase rapidly.

Putting it in context:

- A group of students noticed that weed plants were growing in a previously abandoned field near their school.

- They decided to investigate biodiversity in the field by carrying out a quadrat survey.

- Five quadrats were randomly placed in the field.

- Each quadrat was 50 cm long and 50 cm wide.

- Each quadrat contained 25 smaller squares.

- The total area of the field was 75 m^2.

Results:

Quadrat	Number of squares containing each plant species in each quadrat		
	Meadow grass	Black grass	Ragwort
1	18	23	2
2	5	23	4
3	12	24	8
4	23	25	2
5	17	25	9
Totals			

Design features of the experiment:

1. **State** why the students used five quadrats in their investigation.
2. **Explain** why the students used quadrats of the same size.
3. **Explain** why the students measured the total area of the field.
4. **Explain** how the students were able to identify the different plants.
5. **Explain** why the students counted the number of squares containing the plant species.

Analysis of results:

1. **Complete** the table to show the total number of each type of plant found in the field.
2. **Calculate** the average number of each type of plant found per quadrat.
3. **State** the number of quadrats that would be found in 1 m^2.
4. **Calculate** the average number of each plant species per 1 m^2.
5. **Predict** the abundance of each type of plant in the entire field.
6. **State** the ratio of meadow grass to ragwort found in the field.
7. **Explain** how the results obtained could be made more reliable.

6. Using a transect line

Life on Earth – background and additional knowledge:

- Abiotic factors include light intensity, soil moisture, pH and temperature.

- A transect line is a line across a habitat or part of a habitat. It can be as simple as a string or rope placed in a line on the ground.

- The number of organisms of each species along a transect can be observed and recorded at regular intervals using quadrats.

- A transect line can be used to investigate a change in the distribution of organisms in an area due to environmental (abiotic) factors.

- Transects may be used to investigate how the distribution of species changes from one habitat to another, e.g. woodland to grassland, the top of a slope to the bottom of a slope or from playing fields into natural grassland.

Putting it in context:

- A group of students decided to investigate the effects of abiotic factors on the distribution of plants along a slope.

- A transect line was laid out on the ground starting under the trees, at the top of the slope, and ending at the bottom of the slope.

- The transect line had marks every 1 m. A quadrat was positioned at each of these marks and the number of squares containing each species of plant was recorded.

- The quadrat had 25 squares and measured 25 cm × 25 cm.

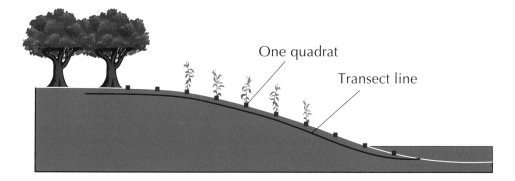

One quadrat

Transect line

Results:

Position of quadrat on transect line	Soil moisture (%)	Average surface light intensity (units)	Number of plants of each type present in each quadrat		
			Plant type 1	Plant type 2	Plant type 3
1	5	1	0	10	25
2	12	3	0	9	25
3	16	40	16	8	22
4	19	50	17	11	19
5	21	60	21	12	17
6	30	70	22	9	14
7	30	80	24	11	11
8	30	90	25	9	9
9	30	90	25	8	6
10	30	90	25	13	3

Design features of the experiment:

1. **Suggest** why the students decided to measure two different abiotic factors.
2. **Suggest** why the students decided to use a transect to place the quadrats.
3. **Suggest** why the transect was laid out randomly down the slope.
4. **Suggest** the reason why the students decided to obtain results from 10 quadrats.
5. **State** the number of quadrats that would be needed to cover an area of 1 m^2.

Analysis of results:

1. **Describe** the trend shown by the soil moisture content from the top of the slope to the bottom of the slope.
2. **Describe** the trend shown by the light intensity from the top of the slope to the bottom of the slope.
3. **Describe** the trend shown by plant type 1 from the top of the slope to the bottom of the slope.
4. **Describe** the trend shown by plant type 2 from the top of the slope to the bottom of the slope.
5. **Describe** the trend shown by plant type 3 from the top of the slope to the bottom of the slope.
6. **Suggest** which type of plant is best adapted to grow in dry conditions. **Explain** how you reached your answer.
7. **Suggest** which type of plant is best adapted to grow in areas of high light intensity. **Explain** how you reached your answer.

7. Measuring the rate of photosynthesis

Life on Earth – background and additional knowledge:

- A limiting factor is one that when it is in short supply the rate of the reaction progresses more slowly.

- Limiting factors in photosynthesis are: carbon dioxide concentration, light intensity and temperature.

- Limiting factors decrease the rate of photosynthesis and so inhibit plant growth.

- Aquatic plants (plants which live in water) release oxygen bubbles when they photosynthesise.

- The bubbles of oxygen can be counted (or collected) and used to compare the rate of photosynthesis.

- A lamp can be used to supply light **but** this will change **two** variables as heat energy will be released. To make sure that the experiment is **valid** the plant will be in water – the water will absorb the heat.

Putting it in context:

- The rate of photosynthesis can be measured using the apparatus shown below.

- The rate of photosynthesis is determined by comparing the rate of oxygen gas production. To determine the rate, the number of bubbles produced per minute was counted.

- The aim of the experiment was to compare the rate of photosynthesis in two different species of water plant – species S and species T.

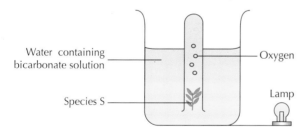

1. **State** which diagram below shows the experimental set-up for species T.

2. **Explain** your answer.

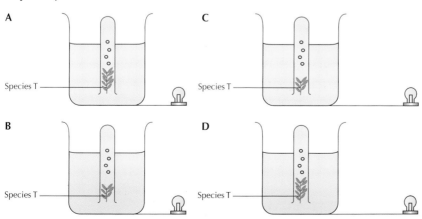

Results:

Time after start of experiment (minutes)	Rate of photosynthesis (bubbles of oxygen per minute)	
	Species S	Species T
1	11	47
2	12	18
3	12	16
4	11	17
5	12	18

Calculate the **average** number of oxygen bubbles produced by species S and species T.

Species S _____

Species T _____

Design features of the experiment:

1. **Give** a variable concerning the lamp that would have to be controlled.

2. **Explain** why the plant had to be the same size for species T as it was for species S.

3. **Explain** why the lamp had to be the same distance from the plant for species T as it was for species S.

4. **Explain** how the students ensured the experimental results were valid.

5. **Compare** the results for species S and species T – which set of results are most reliable. **Explain.**

6. **Suggest** a way to make both sets of results more reliable.

7. **Suggest** why bicarbonate solution was added to all the beakers containing plants.

The above experiment can be altered, as shown in the diagram below, to allow the effect of carbon dioxide concentration and light intensity on the rate of photosynthesis to be investigated.

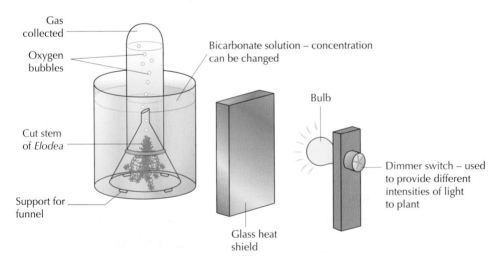

Putting it in context:

- A group of students decided to use the apparatus shown to measure the rate of photosynthesis in an aquatic plant.

- The rate of photosynthesis can be measured by counting the number of bubbles of oxygen released by the plant per minute.

- The aim of the experiment was to find out if light intensity affects the rate of photosynthesis.

Results:

Light intensity (units)	Rate of photosynthesis (number of bubbles per minute)
1	2
2	5
4	15
6	50
8	50
10	50

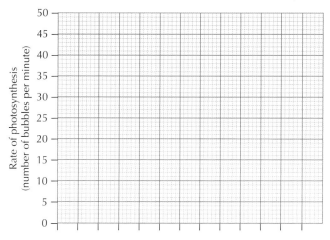

Collect a piece of graph paper and copy the axes on the left.

Use the information in the table above to plot a **line graph** of the results.

Design features of the experiment:

1. **Suggest** a reason for using an aquatic plant in this experiment.
2. **Explain** the purpose of the glass heat shield in this experiment.
3. **Give** two variables that would need to be controlled in this experiment.
4. **Describe** the method used to change the light intensity.
5. **Suggest** a reason why the funnel surrounding the *Elodea* had supports under it.
6. **Suggest** a method that could be used to prove that the gas produced was oxygen.
7. **Explain** why the rate of photosynthesis wasn't measured at 0 units of light intensity.
8. **Suggest** a method to investigate the effect of changing carbon dioxide concentration on the rate of photosynthesis.

Analysis of results:

1. **Calculate** the difference in the rate of photosynthesis at 2 light intensity units compared to 8 light intensity units.

2. **Calculate** the number of times larger the rate of photosynthesis is at 2 light intensity units compared to 10 light intensity units.

3. **Calculate** the percentage increase in the rate of photosynthesis between 2 light intensity units and 6 light intensity units.

4. **Describe** the trend shown by the rate of photosynthesis as the light intensity increased.

5. **State** a conclusion for this experiment.

As part of the research for your **assignment**, you must plan and carry out an experiment to collect your own results.

The "Thinking about Experiments" section above describes some experiments which could be used, or adapted, as a method of generating experimental results which you could use in your assignment. This is something you should consider because you will be familiar with the method and are likely to be able to get some good data. It is not a requirement to use one of these but it certainly worth considering.

The following table offers some suggestions for assignment topic areas where these experiments could provide relevant data.

Thinking about experiments – technique	Potential assignment topic area	Further reading – pages:
1. Measuring enzyme activity	Biological washing powders Food production using enzymes Medicine production using enzymes Converting milk products using lactase Upgrading whey **Catalase activity in immobilised yeast**	54–62, 348–349 SSERC
2. Using a respirometer	Comparing rates of respiration in different organisms Respiration rates in ripening fruits	73–74, 350–352
3. Measuring transpiration using a potometer	Comparing rates of transpiration in plants adapted to different habitats	165–167, 353–355
4. Measuring abiotic factors	**Effects of fertiliser on the growth of algae**	247–252, 356–358, SSERC
5. Measuring the distribution of a species	Comparing biodiversity in different habitats Measuring abundance and random sampling	254–259, 359–360
6. Using a transect line	The abundance of (a named plant) along a transect line Distribution of species across a footpath	254–259, 361–362
7. Measuring the rate of photosynthesis	Photosynthesis and food production **Limiting factors in photosynthesis**	283–285, 363–366 SSERC
SQA exemplar assignment topic ideas	The effect of the concentration of salt solution on the mass of potatoes	33–34
	The effect of temperature on the production of carbon dioxide in yeast	53–62, 348–349
	The effect of temperature during the rising process of dough	53–62, 74–77
	The effect of pH on enzyme activity	53–62

Note – for the potential assignment topic areas shown in bold, more information can also be found on the SSERC website: http://info.sserc.org.uk/biology/biology-national-4/4239-test